Bigfoot Across America

Bigfoot Across America

Philip L. Rife

Writers Club Press
San Jose New York Lincoln Shanghai

Bigfoot Across America

Writers Club Press
an imprint of iUniverse.com, Inc.

For information address:
iUniverse.com, Inc.
620 North 48th Street
Suite 201
Lincoln, NE 68504-3467
www.iuniverse.com

ISBN: 0-595-14314-8

Printed in the United States of America

CONTENTS

INTRODUCTION

Bigfoot—that huge, hairy monster resembling a cross between a man and an ape—is usually thought of as inhabiting the vast forests of the Pacific Northwest. But encounters with such creatures have been reported in 48 of our 50 states (all except Rhode Island and Hawaii). Surprisingly, there are areas outside the Pacific Northwest (among them Chestnut Ridge in western Pennsylvania and sections of the Florida Everglades) where, because sightings are concentrated in a smaller area, the odds of seeing Bigfoot are actually greater. And hardly a week goes by without a new Bigfoot sighting somewhere in the United Sates.

So come along on an eye-opening, coast-to-coast armchair safari in search of America's elusive Bigfoot…

Chapter 1

BIGFOOT IN NEW ENGLAND

Connecticut/Maine/Massachusetts/
New Hampshire/Vermont

One of America's earliest recorded sightings of what may have been a Bigfoot creature occurred in **Vermont** in 1879. Here's how the incident was reported in the pages of the *New York Times* at the time:

Much excitement prevails among the sportsmen of this vicinity over the story that a wild man was seen on Friday by two young men while hunting in the mountains south of Williamstown. The young men describe the creature as being about five feet high, resembling a man in form and movement, but covered all over with bright red hair.

…When first seen, the creature sprang from behind a rocky cliff and started for the woods nearby…Mistaking it for a bear or other wild animal, one of the men fired and, it is thought, wounded it. For with fierce cries of pain and rage, it turned on its assailants, driving them before it at high speed. They lost their guns and ammunition in their flight, and dared not return for fear of encountering the strange being.

There is an old story, told many years ago, of a strange animal frequently seen along the range of the Green Mountains resembling a man in appearance, but so wild that no one could approach it near enough to tell what it was or where it dwells. (1)

More recently, a woman was reading in the woods near East Haven (Essex County) one day in 1976, when she suddenly got the feeling someone or something was watching her. She wasn't prepared for what she saw when she looked up.

Walking towards her was a tall, hairy creature unlike anything she'd ever seen before. Whatever it was, it moved with long, graceful strides. With some relief, the woman realized the creature's path would take it past her position at a distance of about 25 feet at the nearest point.

She later described her visitor as between 8 and 11 feet tall, with a large head and broad shoulders. What little neck there was appeared very muscular. The woman added that the creature had long arms, ape-like hair and human-like eyes. *(2)*

Eight years later (in 1984), a family encountered a similar creature while driving near Colchester (Chittenden County). Here's how the wife later described the incident:

There it was. It just crossed the road and started to go up a hill. It was staring at us with one foot on a wire fence, holding the wire down with its arm.

It was not brown, but white with dirty yellow streaks. His eyes…scared me. They were reflecting a yellow amber color like the yellow on caution lights.

It was at least 10 feet tall. Arms very long. His fur was long. The only flesh part that didn't have fur was his eyes (and) *upper cheeks.*

My husband started to slow down to get a better look. That's when I freaked out and started to hit my husband. I was so scared, and my girls were screaming. So we just kept going. (3)

A couple of years after this incident (in 1986), other Vermont motorists had their own encounters with Bigfoot. In the first, a woman was driving near West Rutland (Rutland County) when a tall, bipedal creature crossed the road in front of her car. She described it as approximately 7 feet tall and covered with dark brown fur similar to a bear. But the witness was certain the creature was no bear. *(4)*

That same evening, a trio of college students were traveling on the same stretch of highway when they nearly collided with a similar creature standing on the road. They described it as between 6-1/2 and 7 feet tall, and covered with black hair. *(5)*

A little over three weeks after this double incident, two other college students were driving between the Rutland County communities of Poultney and Castleton Corners when they came upon a large, hairy creature crouched along the roadway. Both women were positive in their belief that what they saw was not a bear. *(6)*

The hardy inhabitants of Maine have included Bigfoot creatures from time to time in recent years.

In 1973, four children were riding their bicycles near Durham (Androscoggin County) when one of them fell from her bike—and landed practically at the feet of a large "chimpanzee" standing about a yard away on two feet. According to the girl's mother: "All he did was cock his head and look at her." There were no reports of any missing chimpanzees in the area at the time. *(7)*

The evening following this incident, a woman motorist reported seeing an "ape" standing in some bushes in the same area. She described it as being slightly over 5 feet tall and heavily-built, with shaggy black fur. When the woman slowed her car for a closer look, the creature ran into the nearby woods on two legs. *(8)*

In 1988, six campers and a party of Boy Scouts observed a tall, hairy creature they all agreed was neither a man or a bear pulling up roots on Mount Katahdin (Piscataquis County). The witnesses described it as having a dark, triangular-shaped face, broad shoulders and reddish-brown fur. Added one of the scouts: "It stunk like rotten eggs." *(9)*

That same year, a man and his young son had a close encounter with a Bigfoot while hiking in Penobscot County. The man didn't call the creature to his son's attention at the time he saw it for fear of frightening the boy, but later revealed what he'd seen:

…My father had caught movement out of the corner of his eye. When he looked up, he saw what could only have been…a Bigfoot.

…My father described the creature as being at least 8 feet tall, and estimates it weighing over 500 pounds.

The creature had one long arm grasping a tree and one foot up on an incline in the forest floor. (It) remained there staring at my father and I for about 15-20 seconds. My father said it seemed to be bored with us all of a sudden and simply pulled itself up with the one hand grasping the tree and silently disappeared into the woods. (10)

A high school student related the following incident which he said happened as he and his sister were driving near Bucksport (Hancock County) one morning in either 1996 or 1997:

…We both saw something standing there in the road, on two feet, staring at us approaching.

As we approached, we could tell that the creature was about the size of a human. (It) was covered with very dark, shaggy hair, and it was sorta hunched over with its arms hanging down quite far.

It just stood there about a foot or two into the road, watching us. Then, almost as quickly as it had happened, it vanished into the trees. (11)

One of the most recent Bigfoot sightings in Maine occurred in York County in 1999, as described by the witness:

I was walking toward the edge of the wooded area in back of my house…I looked up, and in the woods, behind a few thin trees, was a dark brown creature about 7 feet tall…walking at a normal pace but covering a lot of ground as a result of the large strides he was taking.

He turned his back to me and walked into the woods and disappeared behind some trees. He had only been about 8 feet from me, but I could not see his face well. (12)

Despite their higher human population density, **Connecticut** and **Massachusetts** have also recorded Bigfoot sightings over the years.

One night in 1982, two workers on a dairy farm near Ellington (Tolland County), Connecticut had an experience neither of them will ever forget. They walked around the corner of a barn and nearly bumped into a Bigfoot which was sitting on the edge of a feed bin. Said one of the men: "It was watching the cows, and either playing in or eating the silage." They didn't stick around to determine exactly what the creature was doing. When it suddenly stood on two feet and walked toward them, they decided they'd seen enough and ran off.

In a subsequent report to local police, the pair described their visitor as between 6 and 7 feet tall. They estimated its weight at about 300 pounds, and said its body was covered with dark brown hair. The men added that the creature's nose appeared more like a man's than an ape's. *(13)*

Seven years later (in 1989), a man target shooting at a sportsmen's club near Bristol (Hartford County), Connecticut spotted a large, hairy creature walking on two legs in woods next to the target range. He chose not to use it as a target. *(14)*

In 1997, a man had a harrowing Bigfoot encounter while horseback riding near Thomaston (Litchfield County), Connecticut. Here's his account of the incident:

...There is a stream running through the meadow, and I stopped to let my horse drink...when I noticed the footprint. My first reaction was "Who the heck is walking around here barefoot?" That was when I noticed the size was almost twice the size of my foot.

At the same time, I had a very strange feeling of being watched. The hair was up on the back of my neck. My horse was feeling it, too...She was looking at a hill that rises out of the field...and was very anxious.

Then the smell came. It was foul garbage, garlic, skunk (and) human body odor all rolled into one. At this point, I decided it was time to go.

As I moved Sandy at a canter out of the area...it started down the hill after me. It was huge. I could not get a good look at it, but it sounded like an elephant in full charge. The screams were so loud my ears hurt, and they reverberated in my chest. I could hear branches and trees snapping.

...Sandy bolted, and I stayed on somehow. It kept pace on the other side of a stone wall for about 100 yards, and then backed off. I could still hear the screams for another half-mile. I was beyond terrified. (15)

In neighboring Massachusetts, two men reported seeing a tall creature covered with dark hair and walking on two legs in the town of Mashpee on Cape Cod in the early 1980s. They said the creature was accompanied by a black dog—a circumstance that while not totally unprecedented in Bigfoot annals, is not the typical form of interaction between Bigfoot creatures and dogs (as will be seen repeatedly in other accounts in this book). *(16)*

In 1989, a hiker had a much more detailed sighting of a Bigfoot in a Massachusetts state forest in the Berkshire Mountains, as described by the witness:

...I observed a slight glimpse of a moving object about 100 yards ahead of me...At first, all I could see were arms moving and an occasional glimpse of a fur-covered body...I thought I was looking at a large black bear.

I pulled a pair of 40-power binoculars from my travel pack...At one point, the animal moved into a clearing. It was then that I realized I was observing something very unusual.

The animal was very tall...and slightly stooped. Its body was massive and covered completely with reddish hair...The head was also very unusual in size and shape. The head was rather pointed and covered with hair or fur. The face was dark in color, and had less hair than the top of the head. The neck seemed to be non-existent.

At that moment, it turned toward me, and I was absolutely shocked. It looked very human...I continued to study the animal. It...seemed very occupied. It was moving stones and small pieces of wood and grubbing for either roots or insects...I could see its large arm moving to its mouth every so often. What really took me by surprise was that it was stacking rocks after moving them.

...At this point, I wanted to get a little closer to see if it was a man in a fur coat or a monkey suit...I was able to get about 20 yards closer...I found a good observation area and studied the animal closer.

…I could plainly see that this was something very strange indeed. It was not human. The arms were too long, and the face seemed very distorted and elongated. The arms extended past the kneecaps. The hands were very large, but human in shape. The fur seemed to be tipped with a lighter color of red.

I became so frightened at this point that I began to run back to the trail. I looked back at one point and noticed that the animal had its head back and was slowly moving back into the forest.

…I will never forget what I saw…No way it was a bear or any other forest animal found in this part of the country. (17)

One of the most recent Bigfoot encounters in Massachusetts wasn't as detailed as the one just described, but lasted long enough for the family involved. It occurred in a state forest at the western end of Cape Cod in 1998. One of the witnesses gave the following account of the incident:

…We heard a loud scream. We looked up on the hill and saw Bigfoot!

…It…was about 7-8 feet tall. It started to come down, and my mother gasped very loudly. The creature turned around and walked back up the hill quickly.

I remember the smell it left—awful. (18)

The New England state with the most reported Bigfoot activity in recent years is **New Hampshire**.

In 1971, a woman in Troy (Cheshire County) heard the sound of heavy footsteps outside her house and peered through the kitchen window. She was shocked to see "a really huge guy, about 9 feet tall, in a fur coat" run by and disappear into nearby woods. She later described the fur as honey-colored. The woman's growing suspicions that what she'd seen was not a man were bolstered when she noticed officers of the state fish and game department collecting hair samples and examining broken tree branches along the route taken by the thing. *(19)*

Some four years later (around 1975), two deer hunters encountered a large, two-legged creature near Keene (Cheshire County). Its appearance was preceded by a series of blood-curdling sounds one of the men likened to "a woman screaming in anguish." Then they watched in stunned amazement as the thing ran across a clearing about 40 feet from their position. The men were sure it wasn't a man or a bear. *(20)*

About a year after this incident, a woman had an even scarier nighttime encounter with a similar creature in the same general area. The woman was walking her dog around dusk, when it took off after something and began to bark furiously. The dog later returned with its tail between its legs and clearly frightened.

The woman became frightened herself when the air was suddenly rent by an ear-piercing scream and she heard the sounds of something large moving through the woods towards her.

Keeping perfectly still so as not to give away her position, she watched apprehensively as a two-legged creature at least 9 feet tall walked by only 20 feet from her and disappeared into the darkness. *(21)*

The following year (1977), a man spending the night with his two sons in a camper on the grounds of a flea market in Hollis (Hillsborough County) would have loved to have had 20 feet separating him from the creature he encountered. He was awakened when the vehicle suddenly began to shake:

I opened the door of the truck, threw on some lights and started to step outside, when I saw it face to face. It was all hairy, brown-colored and 8 or 9 feet tall, with long arms (and) *long hair.*

Thank God for the lights, for it apparently startled the creature...It ran towards a fence about 4-1/2 feet high and jumped over it with ease. I could see it standing there in the distance, just looking at us.

I got out of there so fast I left everything I was going to sell the next day behind on the ground. (22)

A decade later (in 1987), a man in Salisbury (Merrimack County) was able to observe a Bigfoot from a safer distance. He was hunting pheasant just after dawn when a strange feeling of being watched came over him. In the words of the witness:

…There he was. Standing right out in the middle of the field. This thing was big. I would say at least 9 feet. Maybe less, maybe more, because I didn't stick around too long to do any measuring.

…The whole body was covered with hair…kind of a grayish color…The face, I couldn't make…out too good…The hands were like yours or mine, only three times bigger…Long legs (and) *long arms.*

It was…like a gorilla, but this here wasn't a gorilla…It would make your hair stand up.

After a few minutes, the creature ran into an adjacent swamp on two legs. The man later reported his encounter to the local game warden, who suggested the creature was probably a bear or a moose. The witness— who'd shot four bears in his lifetime and had never known a moose to walk on two legs—was certain it was neither. *(23)*

In 1998, Bigfoot action in the Granite State returned to the Keene area in the form of a scary early morning encounter experienced by a local taxi driver. The woman was backing up her cab after delivering a passenger to a remote logging road when she bumped into something.

Glancing into the rear view mirror to see what she'd hit, she saw the hairy midsection of a large upright animal she assumed was a bear. That notion was quickly dispelled when the woman turned her head for a closer look.

She found herself staring into the "green eyes of a gigantic, hairy gorilla." When the "gorilla" reached for the car's roof rack, the woman decided this was one fare she wanted no part of and sped away. *(24)*

Chapter 2

BIGFOOT IN THE MID-ATLANTIC STATES

Delaware/Maryland/Pennsylvania/ New Jersey/New York

The term "Bigfoot" wasn't coined until the 1950s. Before that, one of the most common labels used to describe such creatures was "Wild Man of the Woods." Here's how a local newspaper in western **New York** reported on what may have been a Bigfoot creature seen near the shore of Lake Ontario in 1818:

...In the vicinity of Ellisburgh was seen...by a gentleman of unquestionable veracity an animal resembling the Wild Man of the Woods.

It is stated that he came from the woods within a few rods of this gentleman—that he stood and looked at him and then took his flight in a direction which gave a perfect view of him for some time.

He is described as bending forward when running—hairy, and the heel of his foot narrow, spreading at the toes.

Hundreds of persons have been in pursuit for several days, but nothing further is seen or heard of him. (25)

In 1931, there were reports of what may have been a young Bigfoot on Long Island, less than 10 miles from New York City. The episode began when a teenage girl burst breathlessly into the showroom of a plant nursery near Mineola, excitedly informing anyone who'd listen that she'd just seen a broad, hairy ape-like creature about 5 feet tall emerge from the underbrush and walk (on two legs) among the plants. Several male customers grabbed garden tools and advanced toward what one man called the "gorilla." The creature, apparently not liking the unfavorable odds, retreated into the nearby woods.

But the creature wasn't through with its tour of the New York suburbs. That night, there were several more reports of its appearance elsewhere in Nassau County. One woman engaged in a brief staring contest with it in her backyard, and a man said his dog attacked a dark, erect figure that "smelled like decaying flesh." After several more, less-dramatic encounters in the days that followed, the creature dropped out of sight and didn't return. *(26)*

In 1969, three people staying in a cabin at Long Lake in Hamilton County were visited by a voyeuristic Bigfoot. The incident began when one of the women informed her husband that there was a "raccoon" looking in one of the cabin windows.

When the man investigated, he discovered that the "racccon" was in reality the head of a much taller creature. He described the head as large and conical with a dark, flat face and said it was covered almost completely with brown fur. The creature departed when it noticed the man observing it. *(27)*

In 1977, a Bigfoot was observed from a distance of 25 feet by a local policeman and a New York state trooper near the town of Whitehall (Washington County). The creature—a hairy biped estimated to be between 7 and 8 feet tall and weighing between 300 and 400 pounds—covered its eyes when the trooper shone a light at it. The creature then ran off, emitting a sound described as "a loud pig squeal, a woman's scream, or a combination." *(28)*

A dozen years later (in 1989), Bigfoot paid a visit to a home near Whitehall located about four miles from the scene of the two policemen's roadside encounter. This incident began when two men walking in the woods came upon a trail of 20-inch footprints. One of the witnesses tells what happened next:

We decided to head back to his house right away, because we both felt as if we were being watched, and it was close to dark. The entire way home, we felt as if we were being followed.

The next morning, I awoke to see a large creature about 20 feet from the house. It must have been at least 10 feet tall, as the windows are 15 feet off the ground and it was no more than five feet below me. It was brown and looked very human except for its size...and the hair.

After about five minutes, it walked toward the house. As it passed the house, it banged on the wall, and I thought its arm would come right through. (29)

Later that same year, three men reported a harrowing nighttime encounter with a Bigfoot they said chased them near Port Henry (Essex County). Here's how one of the witnesses later described the incident:

...(We) heard a godawful screaming noise...Metal was being ripped and thrown (where) there were...junked out cars. One of my friends uttered the obligatory "What was that?" and the noises stopped.

...The next thing we heard was the sound of something very large breaking through the heavy brush...Before we knew it, it was...standing right in front of us. In the dark, I couldn't see it (well), but I could tell that it was huge. There was a smell...almost like crossing a skunk and deep bog. Nasty.

My buddies took off...I was scared, but sort of curious, too. The thing, whatever it was, was right in front of me. I could feel its breath, hear it breathing. I was only there for a second or two, and then started running.

The thing took off after us. I could hear its feet on the gravel behind me. It stopped chasing us just as we broke out onto the road and headed up the main street.

…Whatever it was, it was big—a lot bigger than a man. And I have never seen a bear move like (the creature). *(30)*

Four years later (in 1993), the main witness in the incident just recounted claimed another thrilling close encounter with a Bigfoot not far from the scene of the previous event. Like the first incident, it took place at night. This time, it occurred while he and a friend were sitting in a parked car between Port Henry and Westport. The man described this second incident as follows:

…I happened to look out the back window, and I could see something coming up behind the car. The LTD wagon had huge taillights and there was no guard over the license plate light, so the back of the car lit the area up nicely.

This thing was big (and) *black, and had its arm out like it was going to open the door on the driver's side…I looked at Matt: "Something's about to open the door."*

Matt slammed the gearshift into drive, and we took off out of there. I looked back…The thing had to be over 7 feet tall. (31)

In 1995, another aggressive Bigfoot chased three deer hunters near Salamanca (Cattaraugus County). In the words of one of the witnesses:

…We thought we saw a bear coming out from behind a tree. But it was running on two feet, and making ungodly noises as it was running after us.

It …nearly caught up to us, but we managed to get to our truck. Before we could leave, though, it smashed out our front window.

…It looked like it was feeding on something behind the tree, when it looked up at us with these yellow, evil eyes. It ran after us yelling like a mix between a gorilla and a man. (32)

Four years later (in 1999), some motorists nearly collided with a jay-walking Bigfoot on a highway farther east along the state's southern tier in Chemung County:

A large (Bigfoot) *ran across a two-lane road in front of our car. We were doing approximately 60 miles per hour, and nearly hit it.* (It) *never looked left or right at oncoming traffic. It was moving very fast, with arms swinging at its sides…We estimate its strides to be approximately six feet.*

We estimate it to be approximately 7 feet tall. (Its) *hair was dark brown or black. The creature was proportional and not fat. I was amazed at its speed and agility.* (33)

Another motorist in the Empire State no doubt wished he'd been traveling 60 instead of stopped along the roadside when he encountered a Bigfoot in Albany County in 2000:

I pulled over to grab a snack…After ripping the wrapper off of the candy bar, I rolled down my window to throw it out. As I did so, I heard this screeching yell that scared the crap out of me.

I opened my truck door to see who or what…made the noise. My first impression was that a group of kids were playing a trick on me. I was ready to start dealing out some harsh words, when suddenly this thing walks in front of my truck's high beams.

It was at least 7 feet tall, and was covered from head to toe with black hair. It stared at me like a deer in headlights until an oncoming car made it move as fast as lightning to the opposite side of the road. (34)

Another state in the Mid-Atlantic region with a history of Bigfoot encounters going back many years is **Maryland.**

In 1909, a farmer was traveling along a road near the town of Lloyds (Dorchester County) when he came upon a grisly scene. The bodies of several cows were lying bloodied in the adjacent pasture. And standing over one of the unfortunate animals was some sort of man-like beast about 7 feet tall. The farmer prudently continued on his way. (35)

Five years later (in 1914), a young brother and sister were searching for a lost cap by lantern light one night near Churchville (Harford County) when they encountered a strange creature seated on a log. It was man-like, but completely covered with hair. The boy later estimated the creature would've been about 7 feet tall when standing.

"It just watched as I walked by," recalled the boy many years later, the incident obviously still vivid in his memory. "I was so close that if it had been vicious, it could have grabbed me." (36)

In 1959, a policeman had a scary encounter with a Bigfoot on a rural road in Carroll County. The officer watched in amazement as the creature crossed directly in front of his patrol car and stepped effortlessly over a barbed wire fence.

He stopped the vehicle and got out, shouting at the creature to stop. The officer had immediate second thoughts about the wisdom of his action when the creature turned around and began to walk toward him. The policeman then drew his service revolver and fired at the Bigfoot. When the bullets appeared to have no effect, the officer sprinted to his car and sped from the scene. *(37)*

In 1973, Bigfoot put in a series of appearances in and around the Carroll County community of Sykesville. One of the best-observed of these began when a man noticed a large, ape-like creature attempt to enter the cellar of a neighbor's house, and then retreat to some bushes in the man's own yard.

The creature then proceeded to play peek-a-boo, periodically raising and lowering its large, egg-shaped head. Apparently tiring of this game eventually, the Bigfoot stood up and walked off on two legs.

The man was able to observe the creature from a distance of 30 feet. He described it as nearly 8 feet tall and grayish in color. *(38)*

A couple of years later (in 1975), a Maryland motorist ran into a Bigfoot—literally—in Harford County. The man was driving his sports car in the early morning hours when the creature suddenly loomed up in the beams of the vehicle's headlights too close to avoid hitting it.

The Bigfoot was stuck by the car's front fender, whereupon it let out a shriek of pain and ran off holding its side. Later, a "bed" of tree branches covered with dried blood was found near the scene of the accident. *(39)*

The following year, a driver had a non-injurious encounter with a Bigfoot on an entrance ramp to the main interstate highway between Baltimore and Washington, D.C.:

I was on my way to work on a Saturday morning about 7 a.m. It was a misty, foggy morning...I witnessed what I thought was a man come up from a steep ravine, walk over to the ramp guardrail and step over it in an angular fashion.

…The "man" was tall, and…a drab kind of mushroom color from head to foot. I thought he was wearing some kind of rain parka or slicker, because there wasn't a clear indentation from his head to the top of his shoulders. As I got closer, I realized that his "suit" was not material, but looked like long, wet, matted fur/hair.

The "man" proceeded to step into the roadway. I thought…this jerk (is) *trying to commit suicide with my help. He got several feet into the lane when he turned to face my oncoming car. He stood there looking perplexed, and…swayed side to side several times as though he didn't know what was coming at him.*

All of a sudden, he startled, turned and ran back to the guardrail. He bolted over the guardrail and went scurrying down the embankment. My mouth was agape, and I thought: "What in the hell was that?" (40)

Later that same year, two Baltimore County policemen came upon a most unusual vagrant while patrolling one evening: a Bigfoot rummaging through garbage cans on the side of the road. They slowed their car and came to within 30 feet of the creature before it sprinted across the road and into nearby woods.

Said one of the officers:

I got a good, clear look at the thing. It was not a bear. It was not a deer. It was definitely on two legs, covered with hair and it moved quick. It was graceful. You could tell it was some kind of animal. I really don't know what it was.

Added his partner:

Whatever it was, it was a giant. That thing was at least 8 feet tall and had to be 500 pounds. This animal was really out there. (41)

In 1990, two sisters living in Dickerson (Montgomery County) were awakened one night by a loud scream coming from outside their house. They looked out a door and saw a "great big, hairy creature" walking across their yard only 30 feet from them. They described it as standing nearly 7 feet tall, and said it walked upright. Six weeks later, a visitor to their home observed a 6-foot-tall creature covered in reddish-brown hair walk through the sisters' yard in broad daylight 40 feet from the house. The man got a good look at the creature's face when it looked in his direction. He said it had reddish areas of skin around the eyes and nose. *(42)*

In 1996, a camper described his encounter with a Bigfoot near Frederick (Frederick County):

I saw it from close up—about 20 yards or so. It appeared to be about 8 feet tall, and its body was completely covered with brown hair. I saw it for about two minutes, until it moved off into the woods...It was a massive creature, bigger than any human I've ever seen. (43)

Although usually thought of as an urban state, **New Jersey** still has considerable areas of uninhabited land. So perhaps it shouldn't be too surprising that Bigfoot-type creatures have put in a number of appearances in the Garden State over the years.

In 1966, a couple living near Lower Bank (Burlington County) discovered 17-inch-long footprints outside their home, and observed a hairy face looking into one of their windows that was more than 7 feet above the ground.

Such developments would have spooked most people, but the couple instead began leaving vegetable scraps outside on a regular basis. These food offerings always disappeared. All went well until one night when they failed to leave anything for their visitor. The result was a garbage can thrown forcefully against the side of their house.

The husband grabbed his rifle and went outside to confront an angry Bigfoot. When a shot over the creature's head produced no results, the man fired directly at the Bigfoot. It wasn't clear if he hit it or not, but the Bigfoot ran off and didn't return again. *(44)*

About four years later (around 1970), a building contractor had a dramatic daytime encounter with a Bigfoot in a wooded area of Sussex County, about 60 miles from New York City.

As the man's truck rounded a curve on a mountain road, the creature suddenly stepped directly into its path. He quickly slammed on his breaks and came to a stop barely six feet away from the Bigfoot. The creature looked at the truck with a startled expression, emitted a loud scream and sprinted into the woods. *(45)*

Around the same time, two other people had a more prolonged Bigfoot sighting in the same area. Their report is of particular significance because one of the witnesses was a game warden.

The pair's attention was drawn to "a loud ruckus" coming from an area of swampland. They weren't prepared for the sight that greeted them when they mounted a small hill that overlooked the swamp. Here's how the game warden described the scene unfolding about 50 yards below them:

(There was) *a huge, hairy, ape-like yet man-like creature standing in about three feet of water. The animal was apparently occupied in a serious dispute with a large, snarling mongrel dog.*

The witnesses watched this confrontation with rapt attention for about 30 minutes. It seemed to them the two animals were fighting over something. Despite the size differential, the dog appeared to be holding its own. The Bigfoot roared, bellowed and swung its arms, but never connected with the dog, which would periodically lunge at it without getting in range of the Bigfoot's reach.

Eventually, the witnesses decided to summon troopers from a nearby state police post. When they returned with the officers, the dog was observed running off, but the Bigfoot was nowhere to be seen. They were able to determine what the two animals had apparently been fighting over, however—a freshly killed deer carcass. *(46)*

A farm family living in a remote area of the same county experienced a series of harrowing nighttime encounters with Bigfoot in 1977. Prior to the first incident, the family's cows uncharacteristically refused to enter their pasture. While the farm wife was trying to herd them into the field, she heard a sound coming from an adjacent swamp that reminded her of "a woman screaming while she was being killed."

Next, she discovered that someone or something had broken into the family's rabbit cages and killed seven of the unfortunate animals. The remaining two were missing.

That same evening, all eight family members observed a 7-foot Bigfoot standing under a mercury vapor lamp in their yard. The wife later provided this description of their visitor:

It was big and hairy. It was brown. It looked like a human with a beard and mustache. It had no neck. It looked like its head was just sitting on its shoulders. It had big, red, glowing eyes.

The family's 70-pound dog lunged for the creature, and the Bigfoot took a swipe at it with its arm, sending the dog flying through the air about 20 feet. The dog promptly ran off, and the Bigfoot turned and walked away. But not for long.

When the Bigfoot appeared under the yard light the following night, the farm couple and two friends were waiting with rifles and shotguns. They all began firing at the creature—and were certain they scored some hits—causing the Bigfoot to growl and run away.

Obviously not knowing when it wasn't welcome, the Bigfoot returned four nights later. This time, the husband tried to run it over with his pickup truck, but only succeeded in chasing it off. The Bigfoot must've finally taken the hint, though, because that was the last the family saw of it. *(47)*

Two years later (in 1979), a boy riding his bicycle through some woods in Manahawkin (Ocean County) had a memorable daylight encounter with Bigfoot. Here's how he later described the incident:

I heard a branch snap no more than 40 feet away. It looked right at me with large, black eyes. It was about 8-9 feet tall...with blackish-brown, rust-colored hair...The hairs on his arm were about 3-4 inches long...He was proportioned the same way a human being would be.

...He just stared...It looked like he stooped his head down slightly and peered hard at me. Then he turned, raised his arm to gain his balance and stepped into the woods...His strides were long. (48)

In 1988, several people got a much longer look at a Bigfoot they encountered in the pine barrens near Mays landing (Atlantic County). According to one of the witnesses: "He stayed in one place looking at us for almost 20 minutes...maybe 25-30 feet away from us." The woman estimated the creature was between 8 and 9 feet tall, and thought it must have weighed around 300 pounds. She added that it had red eyes, and said "the stench from this creature lasted for hours, even through heavy rain." *(49)*

The Mid-Atlantic state with the most extensive modern Bigfoot activity is **Pennsylvania**, where hardly a year goes by without one or more reported sightings of the creature. What follows is a sampler of some of the most interesting Bigfoot encounters in the Keystone State from the past half-century.

Around 1955, a man on a cookout with his four children stared down a larcenous Bigfoot in a face-to-face encounter in rural Wayne County. The incident began when the man heard his children scream, and he rushed outside carrying a friend's shotgun he'd been examining. As he rounded the corner of the building, the man nearly collided with the creature:

...I was so close that I could have touched him. Stunned by the sight of the 8-foot-tall, black, hairy beast, I felt no fear—only curiosity.

The creature was leaning toward the grill, picking up the hot dogs. His arm was extended, and the fingers were stubby and short. His well-formed body would be the envy of any man. He was covered by short, tightly-curled hair.

Trained as an infantryman during the war, I raised the shotgun and yelled, "What the hell are you doing?"

It was only then that he turned to look at me. We made eye contact. Still holding the hot dogs, he had a look of astonishment. I guess he was just as surprised as I was.

We just stared at each other. His eyes were big and brown like a doe's. His nose was slender and well-formed. The face beneath his eyes and around his nose was clear of hair, and the skin appeared to be white.

As he straightened, he looked at the shotgun. I had a feeling he knew it was a weapon. Quick as a wink, he ran off and out of sight.

My kids made a dash for the car. It was only after I stopped for ice cream that they calmed down. (50)

In 1973, a young couple, along with their four-year-old son and the family dog, were sitting in their parked car on the grounds of a church camp near Jumonville (Fayette County) when they experienced a terrifying encounter with Bigfoot. In the words of the wife:

My husband thought he heard something, so he turned the car lights on. There it was. A huge creature about 8 feet tall, with long, dark, matted hair from head to toe. As soon as my husband turned the lights on, the creature slid down over a cliff.

If the family thought they'd seen the last of their scary visitor, they soon learned otherwise. The first indication came from their dog, who suddenly began to bark loudly. The woman tells what happened next:

Our German shepherd is attack-trained, but is usually very gentle. But he jumped toward the back of the car and started barking at the rear window…The dog started biting at the window as though he was trying to grab something.

…When I looked through the window, I could see a pair of red eyes peering at me, and a huge pair of hairy hands clawing at the window. Then the car started shaking.

At last deciding they'd had enough excitement for one evening, the husband quickly started the engine and drove off. *(51)*

A couple of months later, a Bigfoot practically laid siege to the home of a family near the western Pennsylvania town of Indiana (in Indiana County). The episode began when the family's son, standing in the front yard, caught sight of an 8-foot-tall, hairy creature staring at him from the garage. He raced into the house so fast that he ran through the screen door without bothering to open it first. It took a while for other family members to convince him that what he'd seen was undoubtedly a human intruder.

But the following evening, the young man's sister had definite second thoughts about that explanation after she noticed something moving in the trees about 50 feet from the house:

I followed the movement until it came to a clearing. Then I saw it—a huge creature with a slouched, arched back that had long arms dangling at its side. What struck me...was its tremendous stride. I would say that one of its steps would be equal to three human steps. It was no time before it was across the road and up into the woods.

This time, the family reported the sighting to the state police, but the officer they talked with on the phone casually dismissed the incident. Ten days later, the family had much more to tell authorities.

The nighttime ordeal began when the mother and daughter heard an odd thumping sound. The daughter relates what followed:

At first, I thought it was coming from the living room. But as the sound grew louder, I knew it was coming from the back porch.

I switched on the porch light just in time to see a huge form jump from the porch and head toward the front of the house.

The woman rushed to the front door and locked it, at the same time alerting other members of the household to the danger. Her teenage son looked out the front window—and was shocked to see a large, hair-covered head with a flat but man-like face staring back at him.

The boy quickly grabbed his deer rifle and opened the door in time to see the creature retreating toward the road. He got off two shots. He thought he hit the creature, but the Bigfoot continued running into the woods, where it delivered a parting shot of its own—a high-pitched scream.

That's the last the family saw of Bigfoot. The daughter later summed up what she realized others would consider an incredible series of events:

I know that all this sounds like something out of a movie or someone's imagination. People just can't believe that a monster like this is running loose. But I can. I saw it! (52)

Amazingly, these weren't the only dramatic Bigfoot incidents to occur in the state of Pennsylvania in 1973. In fact, on the day before the remarkable series of encounters just described began in Indiana County, a young mother had her own chilling run-in with a Bigfoot in the small Westmoreland County community of Youngstown, less than 30 miles away.

The woman and her toddler daughter had gone to a local cemetery to place flowers on a grave. Simultaneously, she smelled a strong, rotten odor and heard her child cry. Spotting her daughter about 30 feet away, the woman was horrified to see a large Bigfoot walking toward the child.

The woman's maternal instincts kicked into high gear. She raced to her daughter and snatched the child up into her arms—with the creature now only a few feet away—and returned to her car as fast as she could. *(53)*

Early the next moring, another Bigfoot (or perhaps the same one) visited a family in the nearby town of Whitney. A man heard a noise outside his mobile home, and parted the curtains to take a look. Standing no more than five feet away was a Bigfoot. It was looking at the mobile home next to his.

Perhaps thinking his eyes were playing tricks on him, the man called his wife to the window. When she said she saw the same thing he did, they phoned the police. But the Bigfoot left before officers arrived. *(54)*

It was around this same time that two people encountered Bigfoot while walking along some railroad tracks in another part of Westmoreland County. In the words of one of the witnesses:

...We heard some noise coming from the bushes. Just then, this creature crossed the railroad tracks about 50 yards in front of us.

He looked like a big gorilla with long hair on his head, and also had long arms. There was a rotten smell that accompanied it...He was swinging his arms very loosely. (He) *was about 8 feet tall, standing upright on two feet. He also had sort of a humped back.*

...He didn't make any move toward us at all. It appeared that he was afraid of us...We just stood there until he had gone out of sight. (55)

A mobile home park near Elizabethtown (Lancaster County) became the scene of a year-long series of Bigfoot encounters starting in 1975. Events got underway when the park owner installed security lights in response to residents' complaints of someone or something repeatedly pounding on the sides of their homes during the night. The addition of the floodlights soon resulted in sightings to go along with the sounds.

One night, a woman heard a tapping on her window. Thinking it was a neighbor, she turned on her porch light and opened the door. Standing under a light at the end of her driveway was a Bigfoot:

At first, I thought it was someone dressed up, playing a joke. But the longer it stood there, staring without saying anything, the more scared I got...I had this feeling that the longer I stood there, the angrier it got...There was something about the eyes. They seemed to get shinier. (56)✓

The woman quickly retreated inside her home.

Six months after this incident, a young girl looked out the window of her mobile home and spotted a "big bear." Later, she said it looked more like a large monkey. The girl added that the creature was taller than her father and had reddish-brown hair.

In the months that followed, other residents of the park had sightings of their own, including a man who spied a "big gorilla" walking through the grounds. And a woman claimed a similar creature stole her coat. Unfortunately for her, she was wearing the garment at the time. She said the only way she could escape when the thing grabbed her outside her home one night was to wriggle out of the coat and flee indoors. (57)✓

While these incidents were unfolding at the mobile home park, other Bigfoot encounters were occurring elsewhere in the state. One of these was reported by a young hunter in Washington County who came upon a Bigfoot which was sitting on the ground eating apples. He fired at the creature with his 20-gauge shotgun.

The Bigfoot's response was to get up and start walking toward the hunter. The man fired twice more at the creature, but the Bigfoot just kept coming. Suddenly deciding it might not have been such a good idea to begin shooting in the first place, the man started running for home as fast as he could.

A panicked glance over his shoulder confirmed the man's worst fears: the Bigfoot was in hot pursuit. The foot race continued until the Bigfoot broke it off just before the winded hunter reached his house.

Most people would have called it quits at that point. Instead, the man called for reinforcements and waded back into the fray. Summoning a friend with a rifle, he returned to the spot of the initial encounter.

To the hunter's surprise, now there were not one but two Bigfoot creatures eating apples. The men got close enough to see that one of the creatures was eyeing them as they approached. When one Bigfoot stood up, the men opened fire. The results were the same as before. The shots had no apparent effect. The men retreated on the double. But this time, the Bigfoot did not follow. *(58)*

The decade of the 1980s saw a series of remarkable Bigfoot encounters in the Chestnut Ridge area of southwestern Pennsylvania.

In 1982, a man walking in the woods became aware of a foul odor that reminded him of a "mixture of rotten eggs, spoiled potatoes and urine." Shortly thereafter, he was brought up short by the sight of a 7-to-8-foot-tall creature standing next to a large rock on the trail ahead of him. The thing was looking directly at the man.

As the man stood in stunned amazement, the Bigfoot began walking toward him. The creature approached to within 10 feet of the man—who was now "scared to death"—before it left the trail and disappeared into the brush and trees.

Despite his fright, the man was able to notice a number of details of the Bigfoot's appearance. It was covered from head to toe with 6-to-8-inch-long dark brown and black hair. Its long arms extended down to its knees and terminated in large hands with five fingers. The creature's head seemed to rest directly on its "broad and husky" shoulders, with no apparent neck. When it walked, it swung its arms like a man, but its strides were much longer and its whole body dipped and rose noticeably. *(59)*

Five years later (in 1987), another hiker had a similar encounter near the community of Gray Station. It began when a large piece of wood suddenly was thrown near the man's feet. He looked up and saw a Bigfoot, which he described as being between 8 and 9 feet tall, with long hair, a large head, broad shoulders and arms which hung past its knees. Man and Bigfoot stared at each other until the latter turned and ran into the woods with a slightly-stooped gait. *(60)*

The following year (1988), it was the turn of a local night fisherman to meet Bigfoot up close and personal, which he described in a letter to the editor of a local newspaper:

...I experienced a close-up sighting of a large, strange, foul-smelling creature that in comparison to an extra large gorilla would make the gorilla look like a chimpanzee.

...I was in the process of lighting my lantern when I heard a scuffling racket...25 feet away from me. I got my...spotlight. I studied the large creature in every detail. It had reddish-brown fur (and) *extra large fiery eyes that glowed orange color like a bear's.*

...When the creature started coming toward me in an unfriendly manner, I hastily made a run for the safety of my car...I plainly saw the creature at very close range in the bright headlights of my car.

...I am an experienced hunter...and trapper with 50 years' experience. This creature was very, very big and very strange and dangerous-looking. It was definitely not a bear or gorilla. It was too big. The size was unbelievably enormous. (61)

The witness later elaborated on his description of the creature he saw that night. He estimated its height at 6-1/2 feet, and said it had broad shoulders, long arms and no visible neck or waist. The face, which had less hair than the rest of the body, was dark and wrinkled, and appeared more human than animal. He said it made a series of sounds like a combination of wheezing, whistling and a pig grunting.

The man also added additional details regarding his escape from the creature. He said it touched his shoulder and back with its hands at one point, and the smell of its breath reminded him of rotten fish.

The witness believed the Bigfoot became agitated by the flashlight being shone in its eyes, and expressed his opinion that it could easily have harmed him if it had wanted to. *(62)*

Less than a year after this incident, another night fisherman was fortunate not to have provoked a similar aggressive reaction from the Bigfoot he encountered in the Chestnut Ridge area. Still, the witness was visibly shaken when he reported the incident to local police.

According to the man, the episode began when he heard the sounds of something moving through the brush nearby and his dog started to growl. He aimed a powerful flashlight in the direction of the noise. Targeted in the light's beam about 100 feet away was a creature standing approximately 7 feet tall. It had broad shoulders and long arms which reached almost to its knees.

When the light struck the creature's eyes, they glowed a dull red color, and the Bigfoot made a series of loud, shrill noises. Then it turned and walked into the woods, displaying long strides and a slouching gait. *(63)*

In the 1990s, Bigfoot activity in Pennsylvania resumed in the central portions of the state.

In 1990, passengers in two vehicles driving along a rural road in Clinton County got an unusually detailed look at a Bigfoot which crossed in front of them. The first driver actually had to slam on his breaks to avoid hitting the creature, which had vaulted over the guardrail at the side of the road by placing one hand on the rail and swinging the rest of its body over, much as a person might do. The man's pickup truck came to a stop within 10 feet of the Bigfoot, which proceeded to casually observe the man as well as the family of four in the car which had now stopped behind the truck.

The five people were just as interested in observing the Bigfoot, and the truck driver later described it in some detail. He said it was between 7 and

8 feet tall and covered with chocolate-brown colored hair which was long, shaggy, matted and appeared to be molting in places.

The creature's eyes were almond-shaped and about the size of a quarter, with dark centers and lighter areas around the outside, and they moved slowly as it surveyed the witnesses.

The eyes were set in a round head atop a neck so short and thick that there almost appeared to be no neck at all. The body was extremely muscular, with very long arms.

As the Bigfoot turned to walk away, the witnesses picked up its scent for the first time. They likened it to the stench of rotting garbage.

When the creature reached the far side of the road, it turned and made a pig-like grunting sound before disappearing into the woods. The Bigfoot remained on two feet the entire time. *(64)*

Approximately a year later, two young men were standing outside their parked vehicle on a forestry road in nearby Union County when they, too, got an opportunity to observe a Bigfoot at close range.

The witnesses first heard sounds coming from the woods on one side of the road and watched a deer run across. Close on the animal's heels was what they first took to be a "tall jogger." But when the figure stopped on the road about 30 feet from them, they realized they were in the presence of something that wasn't human.

It was between 7-1/2 and 8 feet tall, and they estimated its weight at about 450 pounds. It had long arms which reached nearly to its knees, and was covered with long, stringy brown hair. The facial hair was shorter, allowing them to observe a pair of large, black eyes. The nose was somewhat flattened. No ears were visible.

The wind direction must have changed at that point, because they suddenly noticed a "sour, sweaty" odor emanating from the creature which they hadn't noticed initially.

The Bigfoot appeared to be out of breath. It stared at the two men for about a minute. It moved its thin lips from time to time, but no sounds were heard.

When the Bigfoot suddenly took a step in the direction of the witnesses, the men quickly entered their vehicle. They watched as the creature turned and first walked, then ran into the woods. *(65)*

Six years later (in 1997), two hikers encountered a Bigfoot near the small community of Nisbet in Lycoming County, which abuts the counties in which the previous two encounters occurred:

We were just coming around a bend in the creek when we saw something move behind a fallen pine tree. We then proceeded to get a better view to what we thought was a bear or a deer.

…It seemed to be kneeling down near the base of the fallen tree. It must have heard our approach, because it stood up and…ran into the brush. The only time we got a good look at the creature was when it turned around to run.

…(It had) a sort of man-like head, but with a full brown-haired body like a gorilla. It must have been 7 feet tall, with long arms. The air had a foul, rotten-like smell to it. (66)

Later that same year, a family of Bigfoot provided an indelible memory for a bow hunter in Adams County in the southern part of the state:

I was 25 feet up in my stand…I heard several deer crashing through the laurel towards me…They appeared to be spooked by something…I heard more crunching coming towards me…As it closed in, I saw several flashes of brown hair through the thick laurel.

And then I saw it. It was no deer. It was a bipedal creature about 7-8 feet tall. As it ran by me (it never stopped), I noticed it had matted hair all over it

…Then I heard more sounds coming toward me. Two more bigfoots were approaching very fast. These stopped, and I got a fairly good look at them.

One appeared to be substantially smaller than the first one, and it was dragging a baby bigfoot along with it…It seemed as though they were following the largest one…I noticed that the bigfoots left a foul, garbage-like odor behind. (67)

Despite limited protective cover for wildlife, even tiny **Delaware** has been the scene of occasional Bigfoot sightings over the years.

In the 1960s, police received reports from a number of local residents describing sightings of a "huge, hairy monster" in a swamp near the town of Gumboro (Sussex County)..

In the 1970s, a woman living in Delmar (Sussex County) was said to have seen a white, furry "abominable snowman" in her back yard, and promptly exited through the front door. When she returned home later, she discovered a large knife missing from her kitchen and her dog stabbed to death in the back yard. (Bigfoot may not be to blame for the last act, however. They're not known to enter people's homes and don't require weapons to fend off dogs.) *(68)*

Chapter 3

BIGFOOT IN THE MID-SOUTH

Kentucky/North Carolina/Tennessee/ Virginia/West Virginia

What may have been the earliest recorded Bigfoot sighting in the Mid-South occurred near Wadesboro (Anson County), **North Carolina** in 1956, when a watermelon grower spotted something more sinister than the local kids pilfering his fruit:

It was foggy when I drove up to my riverside field. When I got closer, I could see that the big object appeared to be a man in a stooped over position. Then the creature—the beast—came out of the watermelon patch and disappeared. (69)

More recent Bigfoot encounters in the Tarheel State include the 1982 experience of a hunter near Elizabeth City and the Great Dismal Swamp:

My sighting began with a bad smell. It smelled worse than a dog that had been playing in a sewer. Then I heard it before I saw it, snapping twigs as it walked.

I only saw it from the waist up, from a distance of about 50 yards...It stood in one spot for a good five minutes, eating leaves, so I got a good look at it.

Having seen a bear earlier that day, there is absolutely no way this could be a case of mistaken identity.

This creature had hands like a human that plucked the leaves it was eating, and it looked very fussy about the ones it wanted.

…It looked to be about 8 foot tall, and just unbelievably massive in the chest and shoulder areas. Its fur was very dark black in color.

I only caught a glimpse of its face, as its back was to me…It looked hairy except around the eyes.

The neck was short and stubby, with a kind of fur ridge that ran up the back of the neck to the top of what looked like a slightly-pointed head that sloped down to the forehead like a gorilla. (70)

Five years later (in 1987), a motorist had a low-speed collision with a Bigfoot that he encountered on a highway one night in Onslow County—and some anxious moments following:

…I rounded a corner, and there in the road stood this large, hairy thing. I slammed on brakes and swerved around it, but I must have glanced it, because it bounced over the passenger side of my car.

I slid the car around and stopped where my headlights were about halfway on the thing…I was about 25 feet from the thing, and the smell was horrible.

…It was about 7-to-7-1/2 feet high, (with) reddish-brown, matted-looking hair. Its face was hairless.

…It got up and came at me…He looked ticked…I put the car in reverse, backed up, threw the car in drive and took off. As I was spinning around, I looked in the rearview mirror, and it was right on my car's rear end. I heard a loud bang. About that time, my car caught traction and took off. I left the creature running down the road after me. I have never been so scared in all my life.

After I got to my destination, I checked my car. The passenger rear quarter panel was dented in, with hair on it. And my rear spoiler was shattered where he had hit the rear of my car. I told my friends, family, police and insurance that I had hit a stray cow.

...It took weeks to get that smell off of the outside of my car, and the inside still had that smell. It was ...like sewage and something else.

...I have not been down that road since. I will drive...50 miles out of my way not to go that route. (71)

In 1994, another North Carolina motorist had a less confrontational encounter with a Bigfoot on a road between the towns of Oakboro and Aquadale in Stanly County:

...I saw something standing on the...side of the road...I threw my headlights on bright, and turned the wheel and drove straight to it...As I neared it, I stopped, and I had a clear view of it.

It was so tall that I had to lean closer to the steering wheel to look up into its face. I could see him —clearly a male—from just above his knees to the top of his head.

...He was at least 7 feet tall...with silver and black hair, some of it at least 4-to-5 inches long, and dark, outdoors-type skin, where it was exposed.

His eyes resembled that of a dog's and were the color of an Alaskan husky. (He had a) *wide, flat nose and a wide, thin mouth like a man's. He never opened his mouth.*

He just stood there and looked at me and blinked some and made gestures with his head and eyes. When he nodded at me, I knew it was time to go. So I turned the steering wheel...and drove on down the road. I estimated the time I sat there about 15-to-20 minutes...I never was afraid. (72)

Another state in the region with lengthy Bigfoot traditions is **Kentucky**.

In 1962, a farmer in Trimble County reported that one mauled two of his dogs. He described the attacker as about 6 feet tall and covered in black hair, with long arms that extended down to its knees. Another area farmer claimed a similar-looking creature killed one of his calves with a blow to the head. *(73)*

A decade later (in 1972), a bow hunter had a scary nighttime encounter with Bigfoot near Russellville (Logan County):

I had just climbed down from my…hunting stand and …walked to the edge of a large soybean field. The moon was full, and I thought perhaps I would see deer. What I saw still creates goose bumps.

Standing 30 feet from me…was an extremely large creature. It was between 7 and 8 fee tall, and could have weighed as much as 500 pounds. It had a very short neck and arms that reached nearly to the knees…The face was not visible due to the time of day.

…I didn't shine the flashlight at it as I was too frightened…I cannot detail the fear I experienced getting back to my Jeep…There was no mistake in what I saw. (74)

An Ohio County man was equally certain it was a Bigfoot that watched him as he worked on his property one day in 1976:

I had gone to the back of our farm to get a part off an old junk car. My German shepherd…was with me…It was all grown up around (the car), *so the easiest way to get to the part I needed was to climb over the top of it.*

I heard some noises while I was working, but I figured it was the dog or a rabbit rustling the leaves nearby. When I finished, I was climbing back over the car and heard the noise again…About 20 feet away, behind a small cedar tree, was something big, black and hairy. Unfortunately, I didn't stick around for a good look, but this is what I remember.

Whatever it was, it was standing on two legs…It was watching me through the branches…The hair was black or dark color, with lighter patches that could have been mud or dirt. The arms were at its side.

…I left as quickly as possible, calling for the dog as I ran. He had always been very protective of the whole family, so I thought he would protect me if this thing came after me. He passed me with his tail tucked like he was scared bad. Up till then, I didn't think he was scared of anything. (75)

A Bigfoot was also apparently responsible for spooking some horses during a 1978 incident near Lockport (Henry County):

As we were eating supper, we could hear the horses down at the barn making a lot of noise…My brother and I went down the trail that led to the…barn. I will never forget what we saw.

A very tall, hairy creature was standing by the hitching post…The hitching post only came to its knees, and the hands were hidden by the log at the top…It was very tall and covered with hair.

…I…watched as it walked off…The thing that struck me most was the long strides it took, and the way it swung its arms—very smooth and graceful. (76)

Two years later (in 1980), Bigfoot took a fancy to a family's pet rooster near the Mason County community of Mayslick. The incident began when the father went to investigate a disturbance on the front porch and saw a large, upright creature grasping the unfortunate fowl by the neck. He fired at the intruder with a 22-caliber pistol, but it ran off seemingly unaffected.

The man later described the creature as "man-like" and said it was about 7 feet tall, with long, white hair and pink eyes. He thought it must have weighed about 400 pounds. *(77)*

Several days later, Bigfoot was back in action with another porch invasion in the adjacent Fleming County community of Fairview, as described by the home's owner in a local newspaper story:

At first, I just thought that a dog had gotten into the porch, but then it knocked something over. That's when I grabbed my pistol…and ran around to the back…I saw what looked like a big man running towards the woods. I emptied my pistol as it ran.

…It was very dark, so I really couldn't tell exactly what it was. It stood up like a man, and it was big…It made a thumping noise as it ran.

When the local sheriff responded to the family's call for help, he discovered the intruder's likely target: frozen meat from a freezer on the back porch was strewn on the ground. Other clues were found the next day— 14-by-6 inch bare footprints and long white hairs on the freezer. *(78)*

In 1993, two women motorists got a close-up view of the Bigfoot they encountered on a road between the towns of Wickliffe and Mayfield in the southwest corner of the Bluegrass State, as later recounted by one of the witnesses:

My sister and I…observed a large object on the side of the road. As the headlights shined on…it…we were both stunned at what we saw before us. My sister was driving, and I was in the passenger seat. The creature was on my side, and as the car passed it, I couldn't have been more than two-to-three feet from its body.

(It) was a large, man-like creature with long, shiny, dark hair covering its entire body. With very distinctive muscle tone definition in its legs and arms, this thing looked very strong.

Its face was what startled me the most, because it had human-like features, especially the eyes. (The eyes) were not like any animal's eyes I'd ever seen before, and I've been a hunter all of my life.

It was at least 8 foot tall, because as we passed it, its upper torso towered well above the top of the vehicle.

The creature was standing half in the ditch and half on the road, as if it was trying to cross. When it looked at us, its eyes followed us as we passed. (79)

Six months after this incident, a hiker on Bear Mountain in Estill County was afforded his own detailed examination of a Bigfoot:

I…saw this dark brownish figure some 250 yards away…At first I thought it a rare sighting of a black bear. But when I took out the binoculars, I couldn't believe what I saw.

It was sitting on a log with its back to me, with a rock in its hand, beating on this log…It had a back that was at least 2-1/2 feet across and covered in short, thick reddish-brown hair.

It looked like a giant, hair-covered man with no neck. With the binoculars, I could see in good detail, so there was no mistake.

The hands looked human, four fingers and a thumb, the skin a dark grayish color. The head was conical, like a lowland gorilla.

Suddenly, this thing must have smelled or sensed me, because it stood up and looked in my direction.

It had a bulged forehead near the eyebrow, large round dark eyes, and a flared, gorilla-like nose…The nose and eye areas were free of hair, the skin a dark gray.

...The creature was huge. I later looked at a tree it was standing near, and it had to be near 7 feet tall.

It stared up the mountain at me, and made a peculiar motion with its head, waving it from side to side, and seemed to hunch over a bit with its arms dangling at its knees. It then turned and dashed off through the thicket. (80)

In 1999, two teenagers came in much closer proximity to the Bigfoot they encountered at Cumberland Lake in Clinton County. The brother and sister were walking on a trail when they came upon an 8-foot-tall creature covered with brown hair standing on two legs. It was shaking a tree.

They stopped in their tracks and observed the thing until it turned and began to walk toward them. The girl immediately ran back in the direction they'd come, but her brother watched in fascination as the creature approached. Finally, when the Bigfoot was only about five feet from him, the boy turned and followed his sister. *(81)*

One of the earliest recorded Bigfoot encounters in **Virginia** occurred in 1965 near Alexandria, only a few miles from downtown Washington, D.C. The incident involved three young men who were camping out near their home. One of the participants describes what happened:

Just as I was about to fall asleep, I heard a rasping sound...I ...thought someone...with severe asthma had invaded our campsite...An unbelievably foul odor hit me...It smelled like something that might have emanated from a dead, rotting carcass.

...I turned my flashlight on and directed the beam to its target. What I saw then was nothing human.

Although I had aimed the flashlight for the intruder's head, I illuminated its...arm instead. ...When the light touched it, it jumped...about a foot, then stood...facing me about 15 feet away.

Slumped over but still standing 7 feet tall, it was covered with long, stringy gray-brown hair...It stood on two legs...Its stomach was large, round and bloated, and its arms ended a few inches below the knees. Its fingers were long and slender.

...Although I could see its head, I was too frightened to shine the light into its face, so I never got a look at its (facial) *features.*

Suddenly, the being moved sideways away from me, then turned and dashed swiftly into the forest. (82)

In 1984, a young man riding in a car through the Jefferson National Forest in Wythe County caught a glimpse of some unexpected local fauna:

I (saw) *this large, black, hairy creature walking in the woods...about 20 or 30 yards...from the road...It just walked along undisturbed by the moving cars. My younger brother looked and saw the creature...as well.*

...The creature was...about 8 feet tall, and weighing upwards of 500 pounds I suppose. It had a very long stride, with arms swaying back and forth, and its head a little bit lowered.

...The hair was jet black (from) *head to toe. I couldn't make out any facial features.* (83)

Two years later (in 1985), Bigfoot was spotted on the other side of the state in Colonial Beach. The incident began when a woman spotted a large, hairy figure run across her yard one night. The area was well-lit by floodlights, so when the woman's mother went outside to investigate, she got a good look at the intruder:

When I saw him, I screamed. I was so scared, I could hardly move.

He was about 8 feet tall. His eyes were glowing until he faced the light. His eyes glowed whitish, but when he faced the light, the glow went away. (84)

More recently, Bigfoot put in a brief appearance in the vicinity of Pearisburg (Giles County) near the border with West Virginia in 1997, as related by one of the witnesses:

I was hiking with my girlfriend...We were in a pretty isolated area, so we were surprised when we heard what sounded like a muffled scream. We began to call out...There was no response, so we kept walking in the...direction of the sound.

All of a sudden, we heard crashing like a large animal running through the underbrush...We saw a large figure in the general shape of a human, except more hunched over and hairier…It ran by us about 10 feet away and ran off down the side of the mountain. (85)

In 2000, a motorist reported hitting a Bigfoot while driving along the border with West Virginia near the town of Bluefield. The man said the creature dashed into the path of his car too quickly for him to avoid striking it with the side of his vehicle. The Bigfoot was apparently not badly hurt, because it was seen running off on two legs. The witness said the creature was about 8 feet tall, and had reddish-colored hair. *(86)*

Bigfoot has put in a number of appearances on the **West Virginia** side of the border over the years as well. In fact, some of the most dramatic hirsute happenings in the Mid-South have occurred in the Mountaineer State.

In 1960, one member of a party of young men camping near the town of Davis was cutting firewood when he felt someone poke him in the side. He turned around, prepared to scold one of his buddies for trying to scare him. What he found managed to throw a giant-size scare into the whole group:

(It was) *a horrible monster…It had two huge eyes that shone like big balls of fire…It stood every bit of 8 feet tall, and had shaggy, long hair all over its body. It just stood and stared at us.*

After what must've seemed an eternity to the young men, the Bigfoot turned and walked off into the woods. *(87)*

Later that same year, a bakery delivery man drove his truck around a curve in a remote area near Hickory Flats (between Braxton and Webster counties) and almost ran smack into a Bigfoot which was standing on the highway.

The man stopped his truck a safe distance away and looked back to convince himself of what he'd seen. The creature was still standing by the side of the road. He later described it as 6 feet tall, with "hair all over its body." *(88)*

Two other men had a similar encounter in adjacent Lewis County in 1976. The driver had to brake hard to avoid hitting a black Bigfoot crouched in the middle of the road. When the creature stood to its full height, it was about 8 feet tall. The pair said the creature observed them intently for about 30 seconds before it walked off into the woods alongside the road. *(89)*

In 1982, five young people were enjoying a pleasant autumn evening on a hilltop in Wayne County near where the states of West Virginia, Kentucky and Ohio meet. One of the witnesses tells what happened:

All of a sudden, we heard the dogs over in the hollow below the hill begin to bark like crazy. Then, just as suddenly, they stopped cold...About one minute after the barking stopped, we heard something moving through the brush just over the hill.

I stepped forward to the fence, and was able to make out the shadow of what I thought was the owner of the barn...where the dogs were at coming up to tell us to leave. I called out to the figure, and it was about then that it stood up and huffed and grunted.

This thing stood about 7 feet tall, and...was covered in dark brown fur. I couldn't make out its facial features.

(One) girl was already screaming and in the car, and my other friends were piling in, too, when I made it to the driver's seat. I started the car and turned on the lights.

...This thing covered its face as if to shield its eyes from the car lights, and quickly hurried behind some trees and thick brush. I still didn't get a look at its face, as I was working feverishly to get the car in gear. Needless to say, we got the hell out of there! (90)

Four years later (in 1986), two turkey hunters claimed they were harassed by a pair of aggressive Bigfoot creatures near the small Hampshire County community of Capon Bridge. The witnesses, a father and son, later recounted their ordeal for a local newspaper:

It was overcast, drizzling rain and foggy. We could only see about 50-75 feet...We heard movement...beyond where we could see. We thought it was a flock of turkeys.

But then something strange happened. It sounded as though a turkey flock had split up, and part of the flock was circling to our left and part were circling to our right. Then I heard some branches break....I knew it couldn't be turkeys.... We heard sounds that sounded closer and kept moving around us in a circle.

...I walked about 40 feet...Suddenly...I saw...what looked like a big, black gorilla about 8 feet tall. It was...about 15 feet away.

...The thing and I stared at each other for a couple of moments, and then...it growled...and lunged at me. I turned and ran back.

...I spun around and saw that the thing had stopped about 15 feet away. We both yelled, screamed and flopped our arms. The thing started moving back and circling.

All of a sudden, my dad screamed...I looked, and there was another of the things behind him, and it had ahold of his shoulder...I grabbed ahold of him and pulled him loose from the thing.

The damn thing had (breasts) *like an old woman. It was as big as the other one...It was covered with black hair, but the* (breasts) *didn't have much hair on them.*

...Every couple of seconds, one of them would growl and lunge toward us. We would yell and flop our arms, and it would stop and back up...I looked down and saw my shotgun on the ground. I grabbed it and fired it into the air. The Bigfoot jumped backward. I shot the gun again. They turned and started to run away.

I kept shooting the gun into the air. We could hear them running away through the brush. They ran away in different directions.

The men explained the reason they never fired their guns directly at the Bigfoot was because the creatures "looked too human." They also provided additional details on the creatures' appearance:

They had teeth like a human, but they were much bigger and...yellow....(Their noses were) *a cross between a man and a gorilla. You could see the nostrils real plain...*(They had) *hands and feet of a giant human, except they were hairy on the tops. Their palms and soles of their feet looked like the dark skin of a white man. Their feet were broad and about 18-20 inches long. (91)*

The state in the Mid-South that's seen the most Bigfoot activity is **Tennessee,** where there've been periodic encounters since at least the 1950s. But it's in more recent years that the Volunteer State has become a real hotbed of Bigfoot reports.

In 1959, a homeowner in Knox County told sheriff's deputies he'd seen a "hulking thing about 10 feet tall" looking in one of his windows. The man said the creature ran off when he went outside to confront it. The evening's excitement was far from over, however.

The man and a neighbor were in his yard a few minutes later when the thing returned and crashed against his car. The pair ran to the house, and the Bigfoot moved off a short distance.

The homeowner returned with his shotgun and fired two blasts at the creature from a distance of about 50 feet. The Bigfoot retreated into the darkness and didn't return. *(92)*

Another peeping tom incident involving Bigfoot was reported by a young couple in Robertson County one night in 1966 or 1967. The pair was in the throes of passion in a parked car when they were suddenly interrupted by the sound of grunts and heavy breathing from outside.

They were shocked to see the face of a dark brown Bigfoot pressed against the glass of one of the side windows. Then the creature began to scream and struck the roof of the car.

The man quickly started the engine and drove off with the Bigfoot in pursuit. It followed them to within sight of the nearest house. *(93)*

In 1968, a young woman had a terrifying daytime encounter with a Bigfoot on Monteagle Mountain near Chattanooga:

There was a noise in the woods behind me. I also smelled a very strange odor, almost nauseating, as if something had died.

Turning toward the source of the noise, she was horrified to see a large Bigfoot walking toward her:

I was absolutely frozen with fear. The thing was at least 7 feet tall, and must have weighed several hundred pounds. I'll never forget his enormous chest, and those huge arms and legs. His body was completely covered with a blackish-red hair. The face was a mixture of an ape and a human.

…He seemed to be angry and was growling. I thought he would push me off the cliff or something.

Then, he stopped about six feet from where I stood, cocked his head in a quizzical way, and just stared at me. He studied me for a few moments, then seemed to smile, made a little blubbering noise, and walked back into the brush. (94)

Three years later (in 1971), two young girls were picking blackberries in Overton County when they heard a noise behind them and turned to encounter their own Bigfoot.

Standing about 25 feet away was a creature approximately 7 feet tall with long arms and large, dark eyes. Its body was covered with short (1-to-2-inch long) hair, with longer hair on the head. There was little or no hair on the face. It had a large, long mouth and a nose shaped more like a man than a gorilla.

Inspection of the creature ended when the older girl dropped the berry bucket, grabbed the younger girl's hand and both ran home as fast as they could. Neither of them looked back to see what the Bigfoot did. (95)

The following year (1972), a young man sleeping in a car next to his parents' home in the same county got an unexpected wake up call when the vehicle suddenly began rocking violently.

Gazing out the rear window, he found himself looking into the face of a Bigfoot. He described the face as hairy and generally man-like, with glowing red eyes. There was very little hair on the face, but the other parts of the body he could see were covered with long, dark brown or black hair. It had huge shoulders and a broad body.

The creature continued rocking the car by pushing down on the trunk for a little while longer, then turned and walked away. (96)

Four years later (in 1976), a Bigfoot exhibited similar behavior in Flintville (Lincoln County), when a woman reported an incident to the local police in which she said a "giant, hairy monster" had snapped off her car's radio antenna and bounced up and down on the roof while she cowered on the floor. *(97)*

A few days later, a Bigfoot attempted to snatch a four-year-old boy from his yard in the same community. His mother heard him scream, and rushed outside just in time to grab him and carry him inside the house:

It was 7 or 8 feet tall, and seemed to be all covered with hair. It reached out its long, hairy arms toward Gary and came within a few inches of him.

From the safety of the house, the woman watched as the Bigfoot walked off into nearby woods. She said it left behind a foul odor like "dead rats." *(98)*

Five years later (in 1981), a young boy was badly frightened by a Bigfoot while visiting his grandparents' farm, which straddled the border between Monroe and McMinn counties:

I was playing in the garage...Across the road was a wooded area. As I was playing, I felt like someone was staring at me.

I looked up to find what I first thought was a very hairy man looking at me, until I realized it didn't have any clothes on...It was dark brown to black in color. It was about 7 feet tall. Its facial features were similar to a human's.

I became very frightened and dropped to the ground...and rolled...underneath the car. (99)

The following year (1982), a woman had a much closer encounter with a Bigfoot on a farm near Cookeville (Putnam County). She smelled it before she saw it, when she detected a strong odor unlike anything she'd ever experienced before. This was followed by the feeling that she was being watched.

Entering the barn, she saw it. Although the creature was standing in shadow, she could tell that it was tall, hairy and massively-built, with arms extending to its knees. The woman could also discern enough of its facial features to know it wasn't a bear.

When the creature suddenly emitted a low growl and took a few long steps toward her, the woman turned and ran. She didn't look back. *(100)*

Five years later (in 1987), a teenage boy had an equally-unwelcome visit from a Bigfoot one stormy night on a farm near Calhoun in McMinn County:

My dogs had been acting up all night, barking and…trying to get in the house. Suddenly, they went wild—barking and going crazy.

…I saw a figure walk by my door and stop. Whatever it was, it was huge. It was taller than my door, which was 7 feet high. It was very broad and very hairy. It was tan or dark blond. It made these funny whistling/grunting sounds…Then, all of a sudden, my light on the porch went out and I was petrified.

In the lightning, I saw it walking away, so I…decided to open the door for a better look…This odor hit me, gagging me. My dogs had run under the house and wouldn't come out. I could tell they were scared because they were whining.

Then lightning flashed, and…I saw it about 30 feet away, just standing there…looking…toward the house. Then it made the weirdest noise, like a baby crying. Then, it whistled again, and in the distance, on up in the woods, I swear I heard the exact same whistle.

I could hear it crashing through the woods, and guessed it was leaving. I…moved toward…my door, and slipped and fell on some glass…That's when I discovered that my porch light had been knocked out. (101)

In 1988, a man going to visit a relative near Gallatin (Sumner County) had a brief nighttime encounter with a Bigfoot that enabled him to make a confident estimate of the creature's size.

As he approached a gate on the property, his car's headlights illuminated a tall, two-legged creature covered with reddish-blond hair standing next to it, resting one arm on the gatepost. The Bigfoot quickly turned and ran off. Subsequent measurement of the gate revealed the creature had to have been between 8 and 9 feet tall. *(102)*

In 1992, three persons driving near the town of Halls (Lauderdale County) got a good look at a Bigfoot they encountered standing beside the road, as recounted by one of the witnesses:

I saw a creature that was about 7 foot tall, and weighed about 245 pounds. It was completely but lightly covered with dark brown, long hair.

…As…our car lights hit it, I saw it on the…side of the road. It stood there for a moment, observing the oncoming car. I stared at it, and it stared back…As I was looking at its eyes, I noticed a reddish tint.

His features reminded me of a cross between man and ape. It was standing upright on two feet. And like a human, it ran across the road. (103)

Around the same time (either 1992 or 1993), a Bigfoot apparently took a disliking to three men who were camping in a mountainous area of Overton County. At least the trio interpreted the creature's actions as unfriendly.

The episode began when the men heard rustling and a low growling sound coming from the brush just beyond the area illuminated by their lanterns. Suddenly, the noises intensified, and the men decided to seek what they assumed would be the safety of their car.

They could tell something large was pursuing them, and they felt safe when they locked themselves inside the vehicle. Safe, in this case, proved to be a relative term.

As the men hunkered down in the car, they watched in horror as a huge creature covered in white or silver-gray hair rushed up to the rear of the car, grabbed hold of the bumper and proceeded to vigorously rock the vehicle.

When the driver started the engine, the Bigfoot stood up to its full height of at least 9 feet and let out a tremendous scream. The creature was left standing in the dust as the car set a new speed record for descending the steep mountain road. *(104)*

It was also sometime in the early 1990s that a deer hunter claimed he saw a group of four Bigfoot creatures near Gainesboro (Jackson County).

The man heard the sound of footsteps crunching toward him in the dry leaves, and assumed it was other deer hunters. What he saw instead when they walked within a few yards of his concealed position was like nothing he'd ever seen before.

He later described the four creatures as man-like but covered in dark brown fur, and said their height ranged between 7 and 8 feet. He watched them enter a nearby cave, then retreated from the area as fast and quietly as he could. *(105)*

In 1998, a man confined to a wheelchair had a frightening nocturnal encounter with a Bigfoot near the town of Waco (Giles County).

He'd opened the door of his mobile home and, assuming the dark form lying on the porch was his dog, nudged it with his foot. But, to his surprise, the figure stood upright and started walking toward him.

The man prepared to defend himself as best he could from his wheelchair when the creature suddenly stopped and emitted a low growl. Then it just stared at him for 15 or 20 seconds before turning around and walking off into the adjoining woods.

The man later described his visitor as approximately 6 feet tall, about 200 pounds, and covered with dark brown hair everywhere except around its eyes and mouth. The exposed areas of skin were black and leathery. He said it had wide, thin lips and square teeth much like a human's. The witness added that the Bigfoot also seemed to have a small potbelly. *(106)*

A few months after this incident, a Cumberland County resident claimed one of the largest mass Bigfoot sightings on record near the Hiwasse Wildlife Refuge. In a series of separate incidents, he said he counted as many as 15 of the creatures together at one time—and even formed a relationship of sorts with some of them.

The man described the creatures as covered with dark brown fur and ranging in size between 7 and 8 feet tall. One, which was slightly larger and grayer than the others and appeared to be the leader, communicated to the rest with sounds the hunter found difficult to describe.

The witness began leaving apples in the area where he usually spotted the creatures, and claimed they became so used to his presence that one Bigfoot actually approached him and accepted an apple from his hand.

The witness said his contact with the creatures ended when he moved out of the area. *(107)*

Later in 1998, another Tennessee deer hunter encountered a less sociable Bigfoot near Greenbrier (Robertson County):

I began getting this chill, like something...was watching me...I heard something yell. This was like no noise I have ever heard. I can't even begin to describe what it sounded like.

I...began making my way back to my truck...The whole way there was something at the edge of the woods following me. At the time, I could only make out a big, black form which had to stand at least 7 feet tall, if not taller.

I got in my truck and started it. Then came the biggest scare of my life...It was the ugliest and scariest thing I ever saw. It looked kind of like a gorilla, but then again it didn't. The facial features were not at all right, and the thing walked upright, not on its knuckles.

At this point...it let out (a) *bloodcurdling yell. I threw my truck in drive, and was out of there as fast as I could go. (108)*

Chapter 4
BIGFOOT IN THE SOUTHEAST

Alabama/Florida/Georgia/Mississippi/ South Carolina

The state with the longest-running Bigfoot tradition in the Southeast may well be **Alabama.**

One early incident occurred in Jefferson County in 1946, as related recently by a grandchild of the principal witness:

My grandmother…heard this loud, bloodcurdling scream come from down at the foot of the mountain. She never heard anything like it in her life. She said it made her hair stand on end. It would scream every couple of minutes, and each time…she could tell it was coming up the mountain and closer to the house.

She had no telephone and was three miles from the closest neighbor, so she grabbed grandpa's shotgun, locked all the doors and windows, and turned out all the lights but the porch light. She thought she was dealing with a black panther or mountain lion, because the scream sounded like a cross between a cougar and a woman, but deeper and raspier.

She could tell that the creature was circling the house, and…was staying at the edge of the light…The creature screamed again…It sounded like it was right behind the house...Suddenly, she heard a loud thump, as if something big had jumped onto the porch. She could hear something kind of shuffle...towards the door. She looked (and) saw this huge, shaggy creature on two legs walking across the porch. She said it was covered in shaggy hair and had real long, thick arms.

Suddenly, she heard a pickup truck coming...The creature bounded off the porch in two steps and was gone. My aunts and their dates came up and said they had seen a bear run off...My grandmother told them that was no bear.

About that time, it began to scream, and the boys could tell it was in the woods. They got their guns, and with the headlights of the pickup truck, tried to spot it to get a shot at it. It...screamed a couple of times, and then was gone.

The next morning, there was...oily hair found on nails in the porch floor that smelled like rotten eggs, and a few big footprints in the dust...Grandma heard the screams several more times in the middle of the night, but it never came back up the mountain. (109)

A dozen years later (in 1958), an Alabama farmer went to check on his livestock one night during a particularly bad thunderstorm. He had no way of knowing he was about to lose several of his animals to something far more freakish than the weather:

...I saw this man coming out of the barn door. Leastwise, I thought it was a man. But then there was a flash of lightning, and I saw him clearly for a moment. It made my hair stand up on end.

He was big, and his body was all hairy. He wasn't wearing clothes, and he had a face like a gorilla I saw once at a zoo.

Before I could do anything, he hightailed it into the woods with a pig under one arm (and) a couple of chickens under the other. (110)

In the early 1970s, Bigfoot invaded the small Randolph County community of Wadley. In one incident, a local policeman was sitting in his parked car near some woods just outside town one morning when he happened to glance in his rearview mirror and spot what he at first thought was a tall man in dark clothing approaching.

When the figure got to within 20 yards of the officer, he saw that the "man" was actually a creature covered with reddish-brown hair and standing about 7 feet tall. Deciding his pistol might not be enough to stop the creature if it came to that, the policeman drove into town. *(111)*

In another Wadley incident, a man heard "hair-raising" screams and his dogs barking outside his home one night. Peering out a window, he saw a Bigfoot walking by about six feet away, with the man's dogs yapping at its feet. The witness only saw the creature from the back, but said it stood somewhere between 7 and 7-1/2 feet tall and had long arms. *(112)*

Dogs also figured prominently in the most dramatic Bigfoot incident to occur in Wadley that year.

A local farmer was driving outside town in mid-afternoon when he observed a tall creature covered in reddish-brown fur cross the road ahead of him and walk into the adjoining woods. Since the witness had a 12-gauge shotgun and a pack of hunting hounds in his truck at the time, he decided to stop and give chase.

The dogs picked up the scent immediately and went off in hot pursuit of their quarry, barking more excitedly than their owner had ever heard them. By the time the witness caught up to them, the hounds had treed the Bigfoot.

From a distance, the man could see the creature sitting on a limb about six feet up a large oak. Suddenly, the Bigfoot jumped to the ground and entered into a fierce struggle with the dogs.

As the man watched, the Bigfoot hit and kicked at the circling canines. At one point in the melee, the creature picked up a dog and threw it into the midst of the others.

Finally, the Bigfoot let out a mighty yell and the dogs scattered, running pell-mell past the witness in the direction of the truck. The man decided he was too far away for the shotgun to have much effect, so he just watched as the Bigfoot ran off. When he got back to his truck, he found his dogs licking their wounds. *(113)*

A few years later (in either 1978 or 1979), a teenage boy encountered a Bigfoot that may have been trying to catch fish in a creek in Limestone County:

...I noticed something emerge from the swamp...and wade into the water. It was dark in color, black I think...It stood upright. It...waded into the water knee deep.

I did not really pay attention at first, thinking it was a bear. However, I looked again a few seconds later, and...realized it was not a bear.

It was tall—8 or 9 feet—well built, and very hairy...I did not notice a neck, and the head was round-shaped.

The animal proceeded to look both ways. It then bent over at the waist, took its arm and swung it into the water...The animal did this twice. It then stood up and looked straight in my direction. It stood there a few seconds, then turned around and walked back into the swamp just like a person would. (114)

In 1980, another Alabama boy claimed a much more dramatic encounter with a Bigfoot in Lauderdale County. He told his story to a cousin, who provided the following account of the incident:

...He and his dog were hunting, when all of a sudden, a terrible smell came over the woods. Much worse than a decaying animal. He then heard a grunting sound and limbs snapping off trees.

The creature stepped right out in front of him. He described this creature as at least 8 feet tall and very ape-like. The ape screamed at him in a blood-chilling tone. Out of fear, he fired in the air.

He tried to run, but the ape grabbed him and slung him down a ridge. He fired his rifle at the ape as it charged him. The ape then fled, as did he and the scared dog.

I talked him into going to this spot, and should say I thought it was a hoax. I saw many broken branches and the marks from (a) roll down the hill. But the deciding factor for me was the dog. She whined and smelled the air in fear the whole time we were there. (115)

In 1993, a bow hunter was thankful he was in a tree perch and not on the ground when he encountered a Bigfoot in Dallas County:

I started smelling something first. It was definitely different. Sort of musky. It really stank. Then I saw something come walking down the trail real slow. He was walking upright.

It was making these little noises. I couldn't hear it until he came within 15 yards. Kind of a popping sound, but kind of nasal and muffled.

I was scared. I didn't know what it was. I wanted to get out of there. I didn't want to come down with him still there.

The frightened man observed the creature from his perch. He said it was at least 6 feet tall, dark-colored and had legs longer than a bear. When the Bigfoot finally walked from view, the man decided to make a break for his truck. That's when the hunter became the hunted.

He could hear something trailing him through the woods, getting closer by the second. He was too afraid to look back. When it seemed it was right behind him, the man made a mad dash for his truck and drove off. *(116)*

In 1997, a motorist driving between Chatom and Wagarville in Washington County probably wished he'd consumed one less beverage earlier in the day, when he was forced to stop his car to answer the call of nature:

...In the woods next to the road there was a horrible, hairy, man-like creature. It walked toward me and snorted. I was so scared.

It had what looked like a small limb or something in its hand. It came up to (within) three feet of me, grunted and walked off. I ran back to my car and got away. In my rearview mirror, I saw it standing at the edge of the road as I drove away. (117)

Another southeastern state with a long history of Bigfoot sightings is **Georgia**.

In the mid-1950s, a young farm boy in Troup County encountered Bigfoot while going to feed the family hogs. He said it was about 8 feet tall, appeared to weigh between 400 and 500 pounds and had "long, light brownish hair all over its body." Later, large four-toed footprints were found near the hog pen. *(118)*

Around the same time, a state forestry worker said he was assaulted by a Bigfoot while cutting tall grass near the town of Edison. The witness described his attacker as at least 6 feet tall and covered in gray hair, with "tusk-like teeth and pointed ears." He tried fending it off with his scythe, but was forced to make a run for his Jeep. He claimed the Bigfoot ripped his shirt and scratched his arm and shoulder before he managed to drive off. *(119)*

In 1978, three youngsters said they were chased by a Bigfoot in a sparsely-populated mountainous area near Taccoa (Stephens County). In the words of one of the witnesses:

We saw a tall (about 7 feet), bulky, lumbering figure coming towards us from about 30-40 feet away. It was upright but hunched over. It had a very large head.

We turned around and ran. It chased us and got to within probably 15 feet of us. It stopped before the edge of the woods. We didn't see where it went, because we ran until we were out of breath.

We agreed it looked like a cross between a man and a bear. (We) didn't get a good look at its facial features, as our main goal was to get the heck out of there. I do recall its face was hairy. I remember a strong, musty animal smell. (120)

The following year (1979) a witness got a much better look at the Bigfoot he encountered on the grounds of the Army's Fort Gordon (Richmond County):

(I observed) *a large man-ape creature 10-to-10-1/2 feet tall (estimated weight 1,100 pounds) approaching me with an unhurried pace at a distance of approximately 20 feet.*

The body, except for the face and palms of the hands, (was) covered with short (one inch) dark brown-to-black hair flecked with gray.

…(It had) *dark skin and deep-set eyes. The head sloped back, and was Neanderthal-shaped.* (It had) *no eyebrows, prominent lips (but not protruding),* (and a) *jutting square chin. No ears or teeth were seen.* (It had) *prominent nostrils, (a) slight nose (not flat like a gorilla) and an aged look to the face.*

The body was all big-boned muscle—no body fat—(and it) *had thick arms and legs. It made no vocal sounds, and I did not notice any odor.*

I backed out of the brush and walked quickly back to my car. It did not pursue me.

A bare footprint measuring 22 inches long and 9 inches wide was later found at the scene of the encounter. *(121)*

Four years later (in 1983), a hunter nearly stumbled over a prone Bigfoot he encountered in pre-dawn darkness near the Twiggs County community of Dry Branch:

…Suddenly, something huge stood up beside me about five feet away. This creature screamed at me with a half-roar and half-screaming ear-piercing sound.

This creature stood much taller than me (and) *I'm 6'3".* (It was) *broad from top to bottom.*

It walked off crashing heavily through the brush. It walked about a 100 feet or so and turned and roared again. Then, every 50 yards or so, it seemed to turn and roar back at me. It repeated this four or five times, and that was it.

Whatever this was, it wasn't a human. I must have startled it from its sleep, explaining how I got that close to it. I could see no distinct features other than its massive size.

While all the time holding a loaded (rifle), *there was no way I could've fired. This creature was close enough to have reached out and touched me.* *(122)*

A couple of years after this incident (in 1985), a squad of U.S. Army military policemen had an eventful Bigfoot encounter while guarding the site of a crashed military helicopter and its two dead crewmen in the Chattahoochee Forest, located in the northwest corner of the state. One of the team describes what happened:

Around 10 p.m., I woke to the most godawful howl/scream you could imagine. I could hear metal being thrown around at the crash site.

(Another soldier) *turned on his mag-lite, and what I saw scared the* (expletive) *out of me. Three creatures were there among the wreckage, and they were not bears!*

The closest one—about 10 feet away—was holding a piece of metal from the helicopter. (It) *stood on two legs* (and was) *at least 7-1/2 feet tall.* (It was) *covered in hair except for the face.*

The one behind him was dragging the body of one of the pilots from the wreckage. He (or she) was larger than the first one. My estimate was over 8 feet tall. I only saw the third one briefly, as it was fleeing.

Everyone (went) *into "Dodge City" mode, shooting everywhere. We stayed locked and loaded till sunrise, and did not move from our 360 at the wreckage. Around 6:00 it started getting light, so we moved out to see if we killed one of the creatures. The body which had been dragged from the wreckage was covered with a poncho. We found no creature bodies or blood trails.* (123)

South Carolina was a comparatively late arrival on the Bigfoot scene, but the Gamecock State quickly made up for lost time.

In 1993, a young couple nearly ran into a Bigfoot one night while driving near Pineville (Dorchester County). The man, who was behind the wheel, described what happened:

This thing stepped out in front of us. I hit the brakes. All I saw at first was two long, hairy legs. I started backing up so my lights would spot it better.

This was no bear. This thing was maybe 7 or 8 feet tall and big, hairy and wet. It didn't stand real straight, but more at a slump from the waist up. It had the reddest eyes I have ever seen, and it just looked at us. Frozen, looking, not moving at all. Just standing there.

Then, it just looked to its left and leaned as his leg started to turn, then the other, taking about a three-foot stride until it was out of sight, gone into the woods. This thing walked on two feet and was not a bear. (124)

Three years later (in 1996), a teenager encountered a Bigfoot in a remote part of York County:

I got on my bike and rode up the road. My hound plodded along beside me. My dog noticed it first, and began to slow down.

I looked into the hollow and saw (it). *The creature stood about as tall as a small shed, and was very large. It was covered with a light brown hair. The face was very much like a Mongoloid man.*

It was eating grapes, and its teeth were yellowish. My dog left at this point and went back toward the house.

It looked at me and moved in my direction in a curious motion. I became scared and left. I did look back to see it lumber over the fence. (125)

The following year (1977), the canine residents of the Orangeburg County community of Neeses were the focus of Bigfoot attention in two separate incidents on the same day.

In one episode, a woman who went to see what was causing her dogs to bark soon discovered the source of the problem:

I walked around the dog pen, and there was this Bigfoot pulling on the chain. I started running away, and it started running toward the road. It was just as afraid as I was. It would grunt when it ran.

The witness described the intruder as being 8-to-8-1/2 feet tall, with brown hair. She said its face was human-like, "but more elongated—like a football." It also had large, discolored teeth "like baby blocks." *(126)*

In a second incident occurring around the same time, a nearly identical scenario involving a similar-looking creature was played out at another area home. This time, there were a pair of witnesses, two brothers. Between them, they provided a good description of the Bigfoot they discovered pulling on the wire pen housing the family dogs.

They said it was 8-1/2 feet tall (which they arrived at by comparing it to the pen it stood next to) and clearly a male. It was covered with black hair everywhere except the face and chest. The uncovered skin looked thick and brown. It had large, stained teeth that were "block-shaped" and it smelled like a goat.

As in the other incident, the witnesses and the Bigfoot took off running in opposite directions after their confrontation. One brother said he could hear the creature grunting as it ran. *(127)*

In 1998, a motorist encountered a Bigfoot one night while driving in Williamsburg County between the towns of Andrews and Manning:

I smelled something like a dead animal. The smell was almost like that of gangrene. (Then) I saw what looked like a little fox run across the highway, followed by something big, dark and shaggy. I had to swerve, because I was scared I might hit it as it went across in front of me.

It was about the size of a large man, maybe a little bigger. The creature was on two feet. Its hair (was) *reddish-auburn color. The hair wasn't real long, and it was kind of matted.*

I thought it might be a bear at first, but it seemed very man-like and kind of slender or lanky, not round and thick like a bear. What really freaked me out was its eyes caught and reflected the headlights so they glowed as it went by. (128)

Another South Carolinian reported encountering a Bigfoot along a desolate road in broad daylight near the town of Walhalla (Oconee County). What made the incident more stressful—in addition to the extraordinary size of the creature—was the fact that the witness was on foot at the time:

I saw a gigantic, man-like creature. It was shaking the top of a small tree with one hand. It was 10 feet or more tall. He—it appeared to be male—had fur all over his body, and he had really big teeth. The creature's face was human, not ape-like.

I walked down the road toward the creature, not believing what I was seeing. I got within 45 feet of the thing. It looked at me and smiled the most terrifying smile I have ever seen. I stepped from the road and walked through the woods, afraid to look back. (129)

The flora of the Magnolia State—**Mississippi**—may be fragrant, but some of its fauna share the same malodorous character of Bigfoot creatures found elsewhere.

In 1966, two brothers were driving near the town of Winona (Montgomery County) one night when they had an experience neither of them would ever forget, as later recounted by one of them:

I never in my life saw such a terrifying sight. It was unbelievable, but for real. We were not drinking!

It stood at least 8 feet tall, weighing between 600 and 800 pounds, maybe more. Its shoulders were very straight—at least four feet end to end. Its waist and hips were small—about 40-46 inch waist or thereabouts. It must have been a male.

Its face was a cross between a gorilla's face and a man's face. Its chest was an easy three feet thick. Its legs and arms were huge.

I was less than 20 feet from it. This creature came rushing up to the edge of the highway, and was it ever looking us over. It was so huge and powerful, the thing could have easily smashed the cab of my pickup with one sweep of those powerful arms.

It definitely was not a bear, for I have seen plenty of them in my prospecting days. (130)

In 1975, a Bigfoot provided a vivid childhood memory for some children living in Perlington (Hancock County), as recalled by one of them:

I was asleep in my bedroom one night on the second floor of our home when the neighbor's dog began barking wildly at something in the yard. My siblings and I got out of bed to see what the commotion was all about.

Looking down from our window, we saw something in the garden right next to the house. The thing was only about 20 or 30 feet away. It was squatting down, eating vegetables. I could hear it eating from where I stood. We all screamed. It looked up at us, and kept right on eating.

(It) was built like a human being. It was bigger than the average person. It was covered all over with dark, shaggy brown hair. When it looked up at us, its eyes glowed white.

Eventually, my grandfather woke up and saw it, too. He went out with his gun, and fired into the air. We saw the creature stand up and run off on two legs into the woods. The next morning, we found very large, human-like footprints in the garden. (131)

A few years later (in 1978 or 1979), two people came upon a Bigfoot in the woods of Amite County. In the words of one of the witnesses:

I saw what I first thought was a bear about 25 yards away. It was in a squatting position. It then stood upright. It was, I'm sure, 7-to-8 feet tall.

The creature was covered with very dark brown hair over its entire body, with the exception of its face. It stood upright like a man, not stooped like a bear.

As we departed the area, we watched it follow us. The creature had a long stride, and its arms swung as it walked. It followed us to the edge of the wood line. The only sounds it made were whistling noises. (132)

In 1997, a Mississippi motorist had a memorable nighttime encounter with a Bigfoot on a rural road near the town of Batesville (Panola County):

The creature proceeded to cross the road and came to a halt in front of my car. I viewed the creature for a moment, trying to identify what I was viewing. I then brightened my lights.

This appeared to anger the creature. It proceeded quickly to my car, and bumped into (it), causing the car to move backwards. (The Bigfoot) was leaning on my hood, screaming a very high-pitched sound. I was terrified. I finally managed to get my car into reverse and get around him. I could still hear it screaming in the distance.

The creature seemed to be about 8 feet tall, and may have weighed 400 pounds. (It) was obvious it was a male. This creature had black hair all over, about two or three inches long. The face and chest area were bare. The skin seemed to be a grayish tone. It had a greenish-yellow eye glow. The face was wrinkled, with a broad nose. The head was pointed. (It) appeared to have almost no neck, very broad shoulders and large muscles. (133)

Two years later (in 1999), a Bigfoot played a nocturnal game of cat and mouse with law enforcement officers in a swampy area near Ripley (Tippah County), as told by one of the witnesses:

(I heard) heavy footsteps coming out of the woods. When it got to the edge of the woods, it stopped. I could see the outline of something approximately 7-1/2-to-8 feet tall, approximately 36 inches wide, and very dark hair or fur.

We would take four or five steps, and it would do the same, and stop when we would. We stopped for a moment to try to locate it by sound and by shining our flashlight. We got approximately 20 feet from it at one time, but still couldn't see it.

As we started to move, it bolted. When we turned to see what it looked like, it was already going across the field. We could not see it, but we heard it.

It acted like it was stalking us till we went after it, and showed intelligence when it was making its escape. We also felt it was watching us as we left the field. (134)

Four months after the officers' encounter (in early 2000), a motorist got a better look at a Bigfoot near a swampy area in the same county:

I started down a hill and caught a glimpse of something crossing the road in front of me. I thought it was a deer. When I got to the bottom of the hill, I saw this huge creature standing on the side of the road.

It stood about 8-1/2-to-9 feet tall, (and was) *about four foot across the chest.* (It) *looked like a monkey in the face. The chest area was bare. Over the rest of the body, the hair was roughly four inches in length. It was a dark brown or black in color.*

It threw its arms straight up in the air over its head, and let out a growling scream. I was about 15 feet away from it. I just floored the truck. I might have scared it, too, because it ran into the woods. (135)

The final stop on our Bigfoot tour of the Southeast—**Florida**—is one of the few states outside the Pacific Northwest where your odds of seeing a Bigfoot are actually better than in the creature's more famous stomping grounds of the California-Oregon-Washington corridor. That's because the Sunshine State's Bigfoot encounters are concentrated in a comparatively smaller area—principally in and near the Everglades wilderness.

In 1966, a ranch worker in Osceola County was sitting in a pickup truck in a pasture, waiting for a herd of cattle to come to the gate nearby. Instead, he was approached by a creature no one would want to put their brand on:

Suddenly, this thing growled and came scrambling out of a patch of weeds near my truck. It ran directly toward the truck. I rolled up those windows and snapped the door locks. I started trembling.

As the man cowered in the cab of the truck, the Bigfoot walked up to the driver's side and looked in the window. It growled once more:

It was a low, guttural sound. My mind started working again, and I remembered I was in a truck. I started the engine, and the noise frightened the beast. It moved away. I floor-boarded that truck and roared out of there.

Our faces were only a few inches apart when the thing looked through the window. That was no gorilla. It was more a combination of a human and a gorilla—an apeman. (136)

A number of Floridians have experienced dramatic roadside encounters with Bigfoot. One of these occurred near Brooksville (Hernando County) in the same year of 1966.

A young woman had stopped to change a flat tire on a secondary road late at night, when she became aware of a strong, unpleasant odor and simultaneously heard sounds coming from the nearby woods. Unfortunately for her, sounds weren't the only thing coming from the woods.

A big, hairy creature the witness later estimated to be about 7 feet tall and somewhere between 400 and 600 pounds suddenly emerged from the trees and started walking toward her. It was covered in coarse, black hair, had a peaked head and an ape-like face with large, glowing eyes. It also had a peculiar greenish glow on one side of its body.

As the woman stood frozen in fear, the Bigfoot stopped a few feet away and just stared at her. Then, as if in answer to her prayers, another car approached. The creature made a grunting sound, and ambled off back into the woods. *(137)*

A Nebraska truck driver claimed to have had a much more confrontational encounter with a Bigfoot along another stretch of road in the Brooksville area the following year when he pulled off I-75 to catch a few winks one night. He opened the passenger door of the cab to get more air, and was just starting to doze off when he heard the sound of footsteps in the gravel outside his truck:

I reached over and snapped on the dims. I thought a state patrolman may have pulled up behind me. I started to raise up in the seat when I saw this thing come up to the door of the cab.

This thing was tall, covered with a darkish hair. It stuck its face right into the cab. When the thing leaned in, its body pushed against my feet. That's how close we were. The thing I remember about its face was that the features were pushed in, like a gorilla or like a bulldog. I didn't have time to do any more looking, because these two huge hands grabbed my legs. I fell off the seat and my head hit the gear shift as I was dragged out of the truck.

The panicked trucker now fought desperately to free himself from the creature's vice-like grip:

My blows had no effect. The thing just tucked me under one of his arms like I was a rag doll. He didn't really have a hip like a human. I was just sort of tucked in that big arm. Once, he reached over and sort of rearranged me. My head was pushed down into that fur, and I almost gagged from the stench.

With me under his arm, he walked around to the front of the truck. He moved in front of the lights and looked at me. It was as if he was inspecting me and deciding whether to keep me or throw me away.

I knew I was in bad trouble if he dragged me off. I didn't know what he wanted with me. I just knew I wanted out of there.

I kicked out with my feet against the truck fender. The pressure of my feet pushing against the truck threw him off balance. I could feel him stumbling around. Then, his arm opened, and he dropped me to the ground.

The man wasted no time making a mad dash for the cab of his truck:

He wasn't far behind me, but I had time to slam the door. I hit the door lock, rolled up the window and jumped in the driver's seat. In my excitement to reach the gear shift, I hit the horn button.

The thing went straight up in the air. He came down running. I hit the air horns a few times as he ran into the trees. He didn't look back for a second. (138)

Incredibly, this wasn't the only sensational encounter between a trucker and a Bigfoot in Florida. An incident five years later near Belle Glade (between Lake Okeechobee and the Everglades) may have cured one man forever of stopping to pick up hitchhikers at night. He tells what happened when he spotted a figure along the side of the road:

I figured it was a hitchhiker. I braked my truck, opened the passenger door (and) waited for him to run up.

That sound couldn't have come from a human. I knew there was something wrong. I reached over to slam the door shut. That's when I caught a whiff of this awful stench. It was terrible, like something had died and decayed.

Before he could close the truck door, the "hitchhiker" was standing in the opening:

I don't know how to describe it except as a cross between an ape and a man. The face was human-like. The thing stood about 8 feet tall. The body was covered with dark hair. I almost fainted when the thing stepped up on the running board and started to crawl into the cab.

Deciding the cab wasn't big enough for the both of them, the terrified man promptly exited by way of the door on the driver's side:

I ran like the devil was behind me. After a while, I stopped to get my breath. I saw the thing step down out of the truck and walk back into the swamp. Returning to the truck was the longest, most lonesome 200 yards I've ever walked. I've never been that scared in my life, not even during my tour in Vietnam.

I jumped inside, locked both doors and roared out of there. The stench in the cab was awful. I rolled down the windows, and after a while the smell vanished. (139)

That same year (1971), a team of archeologists discovered something more lively than old bones in the Everglades. In the words of the team's leader:

About three o'clock in the morning, one of the members of our expedition was startled awake by weird sounds coming from just beyond the perimeter of our camp. All of a sudden, there was a thundering noise which sounded as though a wild bull elephant was thrashing about amidst the thick trees and bushes.

The bushes parted violently, and from out of the darkened swamp came a huge creature towering well over 7 feet tall. Its body was covered from head to toe with white hair, and its face looked almost human. The beast walked erect and had long, dangling arms which hung loosely and reached the knee.

It didn't take us long to realize that this was no bear in our midst, no escaped gorilla, nor any other known animal. Our visitor stayed but a few seconds—just long enough for a fleeting glimpse—then it was gone. What remained was a set of enormous tracks in the soft earth and a musty odor which made us gag for an hour after it was gone. (140)

Three years later (in 1974), a motorist reported hitting a jaywalking Bigfoot with his car in the Everglades near the town of Hollywood:

(It) jumped across the guardrail. I just couldn't avoid hitting it. I hit it on the leg with my front fender. It didn't scream or make any sound.

When another motorist reported seeing "a large, 7-to-8-foot-tall hairy thing limping along the highway" near the scene of the accident, a police officer was sent to the area to investigate. He arrived in time to observe the Bigfoot retreating into the brush:

It looked like a man, except it was about 8 feet tall. It was running, beating out a path in the sawgrass. I didn't want to go in after him in the dark. (141)

That same year (1974), a security guard had a shoot out with a Bigfoot one night at a construction site near Palm Beach. He heard something moving in the bushes and aimed the headlights of his vehicle at the spot. The man described what emerged from the bushes and began walking toward him:

It was huge—about 7 feet tall— and was real hairy. (It) was either hunched over or had no neck at all. It smelled like it had taken a bath in rotten eggs. It made my eyes water and my nose fill up.

The witness fired all six rounds from his revolver at the creature from a distance of about 30 feet, and believed he hit it twice:

It grabbed its chest and ran like you can't believe—like a track star. (142)

The following year (1975) two motorists came upon a fellow motorist having a bad night at the hands of a Bigfoot in an area of rural Dade County near Everglades National Park.

The witnesses observed a heavily-built creature resembling a cross between a man and an ape, covered in black fur and standing between 8 and 9 feet tall, rocking a parked car back and forth. As they approached, they saw a man emerge from the car and yell for help. When the head-lights of the witnesses' car shone directly on the creature, it turned and ran off into the mangrove trees. *(143)*

A couple of years after this incident (in 1977), a Baptist minister described a Bigfoot he said he encountered in the Ocala National Forest:

It was standing upright in the middle of some palmetto bushes. (It) was 7-1/2, maybe 8 feet tall. (It) had dark hair on its head and chest, not much on its arms, and none on its face. It had a kind of flat face (and) a flat nose. Its eyes were sunk in its sockets. (144)

In 1983, a hunter related his sighting of a Bigfoot in a sawgrass field near the Big Cypress Swamp:

I saw what at first was just a dark object at the edge of the bayhead. When I was approximately 80-to-100 yards away, it stood up on two feet.

What I saw was no bear. Its back was to me at first, but after about five seconds, it turned to face me. It was dark brown to black in color, stood approximately 6-1/2 feet high, and was very solidly built. I got the height by looking at vegetation where it had stood.

I was able to view the creature for approximately 10 to 15 seconds before it turned and went into the bayhead. I think it took me five or ten minutes to get my heart rate and breathing back to normal. I forgot all about the shotgun I was carrying. (145)

Around the same time (1984 or 1985), two horseback riders encoun-tered a Bigfoot in a swampy wooded area near St. Augustine (St. Johns County):

We noticed a smell like something was dead (and) *we heard branches breaking, like something was in the woods. Then the horses started to act up, blowing, snorting and rearing.*

That's when we saw it. It was about 150 yards from where we just passed, standing in the middle of the railroad tracks. It was in sort of a crouched position, like it saw us and froze, and stared at us.

What I saw was slim and covered in reddish brown hair. It had long arms and was at least 6-1/2 feet tall and around 250 pounds or better, with no neck. It was getting dark, so we couldn't see any detail on the face.

All the time we saw it, we had to fight with the horses to keep them under control. They didn't want any part of whatever it was, and neither did we. It wasn't a bear, and it wasn't a man, unless he was covered in hair from head to toe. (146)

In 1999, a hunter in Osceola County got a bird's-eye view of a Bigfoot that walked past the tree stand where he was sitting. The witness observed a human-like figure covered in black hair walking slightly hunched over. He said it had a cone-shaped head, wide shoulders and no visible waist. He estimated the creature's height at between 8 and 9 feet, and its weight between 400 and 600 pounds. The man also noticed a pungent odor accompanying the Bigfoot that reminded him of the smell in the enclosures of large animals at a zoo. *(147)*

Chapter 5

BIGFOOT IN THE MIDWEST

Illinois/Indiana/Michigan/
Ohio/Wisconsin

One of the earliest recorded encounters in the Midwest involving a Bigfoot-like creature occurred in Gallia County, **Ohio** in 1869. Here's how a contemporary newspaper reported the incident:

Gallipolis is excited over a wild man who is reported to haunt the woods near that city. He goes naked, is covered with hair, is gigantic in height, and "his eyes start from their sockets."

A carriage containing a man and daughter was attacked by him a few days ago. He is said to have bounded at the father, catching him in a grip like that of a vice, hurling him to earth, falling on him and endeavoring to bite and to scratch like a wild animal.

The struggle was long and fearful. Just as he was about to become exhausted from his exertions, his daughter snatched up a rock, and hurling it at the head of her father's would be murderer, was fortunate enough to put an end to the struggle by striking (it) *somewhere around the ear.*

The creature was not stunned, but feeling unequal to further exertion, slowly got up and retired into the copse that skirted the road. (148)

In more recent times, resident's in the Cleveland area told stories of a "big hairy man weighing four hundred pounds" who was said to inhabit a tunnel in Riverside Cemetery for many years until highway construction forced him to move in 1968, at which time he reportedly relocated to a wooded area next to the city's zoo.

As if in confirmation of the creature's new address, a boy claimed he ran into a monstrous something in those woods that very year:

It's 8 feet tall and covered with hair. I chased it, and it knocked me down a slope.

Another youth claimed a similar creature grabbed him in the woods and tore his jacket, inflicting scratches on his shoulder in the process. *(149)*

Before chalking these incidents up to youthful imagination or pranks, consider the fact that four years later (in 1972), a total of eight adults living near the Cleveland zoo reported seeing a black-haired Bigfoot-type creature in the same woods. One man, who was 6 feet tall and weighed 360 pounds, said the thing was bigger than he was. Another witness said the creature stood straighter than a gorilla. (In case you're wondering, none of the zoo's gorillas were missing at the time.) *(150)*

The following year (1973), a cemetery was one of the places local residents spotted a Bigfoot in the Franklin County community of Dublin (near Columbus). The witnesses included two security guards at a golf course who reported encountering an 8-foot-tall hairy something —once standing in a road and another time standing next to a tree. They saw it run off on two feet. Another witness, a local farmer, told police the monster had been scaring his cows at night. *(151)*

Four years later (in 1977), two teenage boys claimed a scary run-in with a Bigfoot in rural Preble County, as later recounted by one of them:

We were walking our dog, and she got excited about something. We smelled this awful stink, like rotten eggs.

We saw this creature that was about 9 feet tall and weighed somewhere around 500 pounds. It had dirty brown hair, and what looked like white eyes. Its arms were real long, and hung almost to the ground.

It chased us across a soybean field towards my home. It seemed like it was right behind us. It took very large steps. When we reached the railroad tracks and the highway, it vanished from our view.

Later, five-toed footprints measuring 14 inches long and seven inches wide were found nearby. The length of the stride between each print was more than six feet. *(152)*

The following year (1978), saw another of those rare reports of a Bigfoot observed in the apparently friendly company of another animal, this time a feline. The incident occurred near the town of Minerva (Stark County), and began when a family noticed what looked like two cougars outside their home one night.

After one of the witnesses directed the headlights of his parked car at the animals, the family was shocked to see a large, hairy, bipedal creature step in front of the felines in what the witnesses perceived as a protective manner. The Bigfoot then started walking toward the car.

At this point, the witnesses beat a hasty retreat indoors and phoned the sheriff's department. In the meantime, the Bigfoot walked up to the house and stood looking in the kitchen window for nearly 10 minutes, illuminated by a porch light. It left before the deputies arrived. *(153)*

A decade later (in 1988), another peeping Bigfoot put in an appearance at a mobile home in Coshocton County, when a woman awoke in the wee hours of the morning to find a large creature covered with black hair and sporting luminous red eyes staring through her bedroom window. The pane the creature was looking through was later found to be 7-1/2 feet off the ground. *(154)*

The following year (1989), a motorist in the same county reported that a Bigfoot crossed the road in front of her car one evening around dusk. She described it as approximately 7 feet tall and covered in black hair, with long arms. The witness first spotted the creature standing by the side of the road, then watched as it traversed the highway in three or four long strides. *(155)*

That same month, two people driving in adjacent Muskingum County got a more detailed look at the Bigfoot they saw standing on a road embankment. One of the witnesses later described it as gorilla-like but standing erect, with long, stringy black hair everywhere except its face, which had black, leathery skin. She also said it had a flat nose, no discernible neck and long arms that extended below its knees. The Bigfoot watched the witnesses' car approach to within 100 feet of where it was standing before it turned, ran up the embankment and disappeared into some woods. *(156)*

A couple of years later (in 1991), two anglers had a close encounter with a Bigfoot while night fishing along the banks of a Muskingum County creek. The incident began when they heard strange sounds coming from the other shore that they likened to a combination of a baby's cry, a man moaning in pain and a high-pitched operatic wail. Whatever was making the sounds kept coming closer, until it seemed like it was on the opposite bank. Then the noises stopped.

A short time later, one of the witnesses got a strike and jumped up to grab his fishing pole. His sudden movement must've startled their visitor, because a large, hairy creature with long arms stood up within the light of their lanterns and walked into the darkness on two legs. *(157)*

In 1996, Bigfoot paid a return visit to Preble County, when a mother and daughter driving on a rural road near the town of Eaton one night stopped their car to let an animal they at first thought was a deer cross in front of them. But when the creature halted less than 10 feet away in the beams of their headlights, the witnesses soon realized this was no deer— nor any other animal they'd ever seen before.

It was hair-covered, stood on two legs, and was so tall that the top of its head extended above the reach of the headlights. It also emitted a stench so disagreeable that they were forced to roll up the car windows.

After staring at the car for a few moments, the Bigfoot turned and walked up a steep slope on the other side of the road with long strides and a gracefulness that impressed both of the witnesses. *(158)*

Another Midwestern state with a long and eventful history of Bigfoot encounters is **Michigan.**

A now-elderly woman recently recounted an incident that happened to her and her family at their home at White Lake near Fenton (Livingston County) in the year 1910:

One day when I was about seven years old, I was standing on the hill lead-ing down to the lake when I saw five of them at the edge of the water. I ran to get my parents. I, my father, my mother and younger brother watched the five of them swim over to the little oak woods on the north side of the lake.

They had no clothes on and they were hairy. There were three large ones and two small ones. We never saw them again. (159)

In 1956, three migrant workers had a traumatic encounter with a Bigfoot one evening near Marshall (Calhoun County). One of the young men came running breathlessly into their quarters to announce he'd just seen a huge, hulking figure outside. His two friends went out and found nothing but a peculiar odor. They started to go back inside when the nightmare began. One of the two tells what happened:

It must have been behind us, for all of a sudden, I felt arms wrapping around me, and I was hoisted off the ground!

The man who'd first spotted the creature was still inside searching for his shotgun when he heard his friend scream. He raced outside and, think-ing quickly, turned on the headlights of a parked car to provide illumina-tion. He was unprepared for the scene which greeted his eyes: a huge, black monster was walking off with one of his friends tucked under each of its large, hairy arms.

The sudden glare of the headlights surprised the Bigfoot, which stum-bled and dropped one of the men. The man immediately charged into the creature, causing it to drop the second man. It then "ambled off in a sort of stumbling, sidewise motion" and disappeared into the nearby woods.

In addition to traumatic memories of their ordeal, the trio was left with an awful "rotten egg" odor on them. They wound up burning the clothes they were wearing during the incident. Comparing notes afterward, the witnesses agreed their would-be abductor had been about 8 feet tall and had "big green eyes." *(160)*

In contrast, the two Bigfoot creatures which a logger encountered in Ontonagon County in the 1950s were downright helpful. The man had been pinned by a tree which fell on him and passed out. When he came to, he found himself being rescued by a pair of Bigfoot creatures. One lifted the tree while the other dragged him free. His benefactors then walked away. *(161)*

It's hard to say what the Bigfoot encountered by a mother and daughter driving in rural Monroe County one night in 1965 actually had in mind, but neither woman was inclined to ascribe a charitable motive to it. The mother told how the incident began:

I saw something. I told Christine not to stop. But it was on us. I screamed and Christine screamed. I looked over at my daughter, and there was this huge, hairy hand on top of her head.

Her daughter recounted events from her perspective:

He reached through the window and grabbed my hair. He was all hairy. And the hairs were like quills. They pricked me whenever I touched them. I think it had jumped on the car and its hand got caught in my hair, and it was simply trying to get loose.

Before the Bigfoot managed to extricate itself, the young woman's head got bumped against the car's door post, and she fainted. When she revived, her arm was covered with what appeared to be saliva and her hair was stiff from the same substance. The two women later estimated that their unwanted hitchhiker was about 7 feet tall. *(162)*

In 1976, two boys (ages 8 and 9) came upon a Bigfoot by the Huron River near Brighton (Livingston County), as recalled years later by one of them:

We noticed an odor that smelled like old roadkill mixed with cabbage. While crossing the bridge, my friend grabbed my arm and jerked it as if to show me something.

About 20 feet below us was the biggest pair of hairy shoulders I had ever seen—as if it was wearing football pads under its skin. Because of this, its head looked almost small.

The creature looked like an ape. It had dark brown fur which covered every part, though we never saw its face.

It was crouching by the bank of the river, looking down into the water.

The Bigfoot was apparently unaware it was being watched, and the two boys decided to keep it that way. They quietly continued on their way. *(163)*

A couple of years after this incident (in 1978), two sheriff's deputies lost a foot race with a Bigfoot in neighboring Shiawassee County. Fortunately for them, they were the chasers, not the chased.

The pair responded to a call from several local residents reporting a large creature observed outside their apartment building. When the officers arrived at the scene, they spotted and gave chase to a "large, upright, furry animal about 8 feet tall" that moved on two legs. They lost their quarry when it ran up a creek and disappeared into thick undergrowth. *(164)*

Three years later (in 1981), a teenage girl living on a farm in Michigan's upper peninsula had a touching experience with a Bigfoot—literally. She described what happened when she and a younger sister entered the family barn one evening:

Roxanne was scared to go into the barn because she'd heard noises. I said I would go first, so I reached for the light. I felt fur. It felt thick and dirty. At first, I thought it was a goat. So I took off my glove and I touched it again.

The girl's eyes must've adjusted better to the darkness by this point, because she now got a better look at what she'd been touching. She didn't like what she saw:

It was tall, with red eyes. (It was) *big and black and furry, and stood on two legs. It had a deep growl.*

Trying desperately to remain calm, the girl instructed her sister to return to the house, while she followed, slowly at first. But, she said, "It started walking out behind me, and I started running."

An older cousin came to the sisters' aid armed with a shotgun. He tells what happened next:

I didn't shoot to kill. I just shot into the air to scare it away.

It was about 6'6" or 7'6". It was standing on two feet, and had real long arms. It was some kind of animal between a bear and an ape. (165)

In 1992, two men driving along a back road near Bay City (Bay County) suddenly found themselves sharing the road with a pair of Bigfoot. The men said the creatures walked out in front of their car, forcing the driver to stop to avoid running into them. Then the Bigfoot began walking toward the car. At that point, the men decided to take a different route to their destination. They quickly turned the car around and sped off.

The witnesses later described the larger of the two creatures as being between 7 and 8 feet tall and weighing an estimated 500-600 pounds. The other Bigfoot was about 5 feet tall and weighed approximately 300-400 pounds. *(166)*

Six years later (in 1998), a man in Ontonagon County found himself perfectly positioned to tell the difference between a bear and a Bigfoot:

I was sitting in a tree stand overlooking a bear bait pile, hoping to see a bear. Well, I did see a bear. While this large bear was eating some fish off the pile, me and him both heard some screeches. When we heard this again, the bear looked over towards a creek.

All of a sudden, this bear took off as if I shot at him. I mean this bear was terrified. I figured a larger bear was in the area. After watching and listening for a while and not seeing anything, I decided to leave before it got dark on me.

Just as I was about to climb down from the tree, I heard a twig snap over by the creek. When I looked, I seen a creature that was walking upright and stood taller than any bears we have in Michigan. That's when I knew it was a Bigfoot-type thing.

I had a firearm on me at the time. However, I would not shoot the creature unless it was attacking me.

As it approached to within 25 yards, it suddenly stopped and looked right at me. It then lowered its head and took off running towards the creek. (167)

Residents of Michigan's neighbor to the west—**Wisconsin**—have had some interesting Bigfoot tales of their own to tell over the years.

In 1964, a man encountered a Bigfoot while driving on a rural road at night near Delavan (Walworth County):

I had the bright lights on and (they) *picked up this big, hairy creature. He jumped the barbed wire fence, about four feet high, with ease, and was heading across the road. It scared the devil out of me.*

It was big, dark and hairy. He was standing up just like a human. He had hands and feet, and was running like a man does, bent just a little bit forward. It was too hairy to see if it had ears. It had long arms. The closest animal it looked like was an ape. It was too big to be human.

The witness later estimated the height of the Bigfoot he'd seen at between 7 and 8 feet tall, with a weight somewhere between 400 and 500 pounds. (168)

In 1970, there were a flurry of Bigfoot sightings in the vicinity of the small community of Benton (Lafayette County). In one incident, a woman and her son were turning their car around in farmer's lane one night when they spotted a creature unlike anything either of them had ever seen before illuminated by the glow of their car's taillights. According to the woman:

I looked back and saw a big, light-colored thing standing about 30 feet away. It looked like a husky, ape-like creature about 6 feet tall. The thing stood with its feet apart, and its arms dangled down to the ground.

A neighbor boy corroborated the woman's story from another vantage point:

I was standing and watching the car turn around. I saw the thing. It was solid fur down to its feet. It took off out of there fast. (169)

A couple of years later (in 1972), a woman living in a rural area near Fort Atkinson (Jefferson County) reported a Bigfoot took its frustration out on her horse when it failed to gain entry into her house.

She told an agent of the state department of natural resources that a "large, ape-like creature" rattled her front door in an apparent attempt to get in. Failing, the creature then walked to a nearby shed where a horse was kept. The horse was later found with a long, fresh gash. It was unclear if the wound was the result of a blow from the Bigfoot or whether the horse had injured itself trying to escape. Bare footprints over 12 inches long were later found in the ground leading away from the shed. *(170)*

Four years later (in 1976), a farmer living in Cashton (Monroe County) suspected Bigfoot of bothering some of his livestock as well. The man observed a dark, hairy, bipedal creature about 7 feet tall on his property. He said it had a "staggering" odor, and emitted a sound "something like a young bull would make."

When the man's dog attacked the creature and bit it on the leg, the creature brushed the dog aside with a sweep of its arm and walked off. The man and his wife noticed saliva on the dog's coat immediately after its encounter with the Bigfoot that smelled identical to the strange odor they'd recently noticed on some of their cows. *(171)*

In 1981, three teenagers claimed they were terrorized by a Bigfoot while staying at a cabin near Plainfield (Waushara County).

The first thing to unnerve the trio was strange sounds coming from some nearby trees:

We heard these deep growls in the woods and couldn't figure out what they were made by. Whatever it was, it yelled about 10 times in a row.

Later, one of them saw what may have been responsible for the sounds:

I had gone down the outside stairs, and when I got to the bottom, I saw it standing next to the car. (It was) at least 7 feet tall, with huge shoulders.

The witness said the creature was completely covered in thick, white hair. The trio caught brief glimpses of the creature in the woods later the same day. Said one:

I could see the upper half of its body. It kind of bent sideways. Then it ducked down and shot out towards the house. I went in and locked the door.

The next day, two of the witnesses heard clawing and pounding on the outside of the cabin. That night, one of them saw two of the creatures outside the cabin. He said one of them started to ascend the outside stairs, but retreated when he yelled at it. One of the creatures was also believed to have broken one of the cabin's ground floor windows.

The Bigfoot became bolder the following night, showing up outside the cabin's second floor living quarters:

It came right up onto the deck. We could see it walking back and forth in front of the sliding windows. We sat there on the couch with axes and hatchets and kitchen utensils for weapons.

Fortunately, the young people never had to use their weapons. The Bigfoot left when daylight arrived, and so did the teenagers. *(172)*

In 1992, two other Wisconsin teens had a briefer but equally memorable encounter with a Bigfoot a few miles from the scene of the 1972 incident involving the scratched horse.

The pair were driving along a rural road on a foggy night when they noticed a "strong, skunky odor" followed by the sight of a strange creature on the side of the road. In the words of one of them:

It was large. Its lower chest and upper belly was at the top of my car. It was hairy, and it was standing over a roadkill. It wasn't a bear.

Apparently, the Bigfoot didn't think kindly about sharing its meal. It lunged at the car, inflicting scratches to the side. The witnesses then quickly drove off. But one of them had another Bigfoot sighting along the same highway a couple of months later.

He was riding with a second friend around dusk when they both observed a creature between 7 and 8 feet tall walking at the edge of a corn field about 20 feet from the road. When the driver slowed for a better look, the Bigfoot looked directly at them. Then it turned and walked off through the corn. The witnesses could see the creature's head and shoulders above the top of the corn, which was about 6 feet high. *(173)*

In 2000, a newspaper deliveryman encountered a Bigfoot on a road near Granton (Clark County):

He was all covered with hair, a real dark gray color, with some spots that looked honey color. It was walking on two legs. And it was mighty big.

You better believe I was scared. That creature could have tipped that car upside down and thrown it in the ditch. It was that big.

The witness came within about 35 feet of the creature. He estimated that it was approximately 8 feet tall and must have weighed about 500 pounds. The Bigfoot looked directly at him at one point, and he said it had an ape-like face. The witness added that the creature appeared to be carrying a small goat or sheep in one of its hands. *(174)*

During the decade of the1970s, Bigfoot paid a number of visits to the state of **Illinois**.

In 1970, a motorist claimed he was attacked by Bigfoot when he stopped his car due to engine trouble near Olive Branch (Alexander County). He was peering under the hood when he heard a noise off to the side.

Almost before the man knew what was happening, he was engaged in a fierce struggle with a hairy monster standing over 6 feet tall. In the intensity of the fight, the witness never got a really good look at his attacker. But he noticed that the creature's hair felt like steel wool.

Just in the nick of time, a large diesel truck drove by, and the noise of its engine apparently frightened the Bigfoot off. The victim caught sight of a large, black figure run on two legs into the adjacent Shawnee National Forest. *(175)*

That same year, two young boys encountered a Bigfoot at a much safer distance along a stream near the town of Colison (Vermilion County), as recounted years later by one of them:

My friend was in the lead when he suddenly stopped. I was about to say something when he motioned for me to get down.

We saw a man-like creature at the water getting a drink. I thought it might be a hunter. (Then) *I noticed that the "clothing" it wore looked more like hair. After we watched this thing for about two minutes, it stood up and looked around, and began staring in our direction. We were about 200 feet from it.*

At this point, I could see it was not a person, but a creature that resembled an ape. It had dark brown, almost black, hair all over it except for the face. Its eyes were a reddish-orange. It was tall—maybe 7 foot or so— and its head was slightly pointed. We did smell a strong odor, like something rotten.

It was slightly bent forward. Because of its posture, I thought it was getting ready to start running after us. But, instead, it turned around and headed away from us into the woods. (176)

In 1972, "two reliable citizens" informed police in the East Peoria area of Tazewell County that they'd observed something 10 feet tall resembling "a cross between an ape and a caveman." One of the witnesses said the thing emitted a strong odor that reminded him of a "musky, wet-down dog." *(177)*

The following year (1973), a young couple parked along the Big Muddy River near Murphrysboro (Jackson County) one evening had their romantic moment interrupted by a most unwelcome visitor.

Alerted by a sound from nearby woods that one of the pair described as "about three times as loud as a bobcat, only deeper," they were suddenly stunned to see the source of the noise—a huge, hairy, bipedal creature about 20 feet away that was walking directly toward their car. The quick look they got at the creature before they sped off revealed an animal about 7 feet tall covered with dirty white hair. *(178)*

The following night, a teenage couple was sitting on a porch near the scene of the previous incident when they heard movement in nearby bushes. What they saw when they looked in the direction of the sound sent chills down their spines.

Standing in an opening in the trees was a dirty white creature about 7 feet tall and appearing to weigh about 350 pounds. It had a round head and long, ape-like arms. But what was most unsettling to the witnesses was the fact that the creature's eyes were glowing without any light source to cause a reflection. They also detected a strong odor coming from the direction of the thing that reminded them of foul river slime.

The boy, who was the son of a state policeman, said later:

It took me 15 minutes to get scared. But then it really hit me.

Eventually, the creature turned and walked off into the darkness. *(179)*

In 1975, a teenage girl glanced out her bedroom window early one morning at her home near Vandalia (Fayette County) and was shocked to see a Bigfoot in the yard:

I saw a very large black thing. It was bent down looking at something in the snow only six feet from my window. It looked up at me and slowly stood up. It was very tall. (Its) *body was huge. The eyes just stared at me. I ran in my brother's room. He was afraid, too, so we hid in his closet.*

The girl later added that the creature had hair everywhere except around its eyes, which she said had a reddish tint to them. The face looked more like a man than a gorilla. *(180)*

Indiana has a record of periodic Bigfoot sightings stretching from the 1960s to the present.

In 1969, a farmer near Rising Sun (Ohio County) was walking with his dog one evening when the animal suddenly started to growl and bark. Looking around, the man spotted the cause of the dog's agitation. Standing about 25 feet from them was a man-sized figure covered in black fur. The witness described what happened next:

I watched it for about two minutes before it saw me. It stood in a fairly upright position, although it was bent over about in the middle of the back, with arms about the same length as a normal human being. I'd say it was about 5'7" or 5'8", and it had a very muscular structure.

The head sat directly on the shoulders. The face was dark black, with hair which stuck out on the back of its head. (It) had eyes set close together, and a very short forehead.

It was all covered with hair except for the back of the hands and the face. The hands looked like normal hands, not claws.

When the witness moved toward a car parked nearby, the Bigfoot emitted a "strange grunting sound." It then turned, leaped over a ditch and ran off on two legs. *(181)*

A couple of years later (in 1971), another farm in the Hoosier State was the scene of a now-familiar tussle between a Bigfoot and a dog. It happened near Sharpsville (Tipton County) and was described by the dog's owner:

One night, all the dogs started barking. I went outside to investigate, and there was my dog lunging at a thing that was standing in a low spot. It was still taller than I was.

The head wasn't shaped like an ape's, and I don't think it looked like a man's head, either. It looked like a helmet, but it was furry. It didn't look right for the ape-looking body it had.

It would swing at the dog kind of like it was slow motion. The dog would run up to it and lunge, but seemed like before it got there, it would hit the ground and jump back. The thing would swing and barely miss it. This went on for two or three minutes.

Finally, the man decided to go into his house for his gun. He returned just in time to see the Bigfoot walking off into the darkness. The witness added a few more details about the creature, including the fact that it was about 9 feet tall, had very long arms, growled in a "deep, rumbling manner" and gave off a "rank and sickening" odor reminiscent of a "decaying meat and vegetable combination." *(182)*

The following year (1972), two farmers near the Putnam County community of Roachdale shot at a Bigfoot that had attacked their chickens. They drove onto their property one night and saw it standing in the doorway of the chicken house, illuminated by the glare of their car's headlights:

This thing completely blocked out the lights inside the chicken house. The door is 6-by-8 feet, and its shoulders came up to the top of the door, up to where the neck should have been. But this thing didn't have a neck.

It looked like an orangutan or a gorilla. It had long hair with kind of a brownish cast to it—sort of rust color. I never saw its eyes or face. It was making a groaning racket.

Then the men and the Bigfoot ran in opposite directions. The men returned and began blasting away at the fleeing creature with their shotguns. Said one of the witnesses:

I shot four times with a pump shotgun. The thing was only about 100 feet away when I started shooting. I must have hit it. I've killed a lot of rabbits at that distance.

There was no indication, however, that the men's firing had any effect on the Bigfoot. *(183).*

The next year (1973), another Indiana man probably wished he'd been hunting instead of fishing when he had his Bigfoot encounter near Kokomo (Howard County). According to the witness:

I saw a large figure 20 feet behind me in the woods at twilight. It was apparently watching me.

Assuming it was another angler, the man resumed his fishing. He thought nothing more of it until a few minutes later, when he felt a hairy hand on his shoulder. The man turned around and came face to face with a tall, hairy ape-like creature standing on two feet. Before he had a chance to panic, the Bigfoot casually turned and walked off into the woods. (184)

In 1977, Bigfoot returned to the Rising Sun area of Ohio County. In the most dramatic incident, a young couple and their baby arrived home one night and found an unwelcome visitor in their driveway.

They'd just gotten out of their car when they spotted a large, hairy ape-like creature approaching it. They quickly got back inside and drove off, but not before the Bigfoot crashed against their vehicle and dented it.

The wife thought their attacker must have been 12 feet tall:

If my husband stood on my shoulders, he'd still have to look up at it. And it wasn't a bear. (185)

Two years later (in 1979), a motorist got a better look at a more moderately-sized Bigfoot he encountered while driving near Bluffton (Wells County):

Around 2 a.m., I saw a large, human-like creature standing alongside the road. The first thing I noticed was the red color of its eyes. My headlights hit those first, I guess, but then I could make out the rest of it as I got closer to where it was standing. It was at least 7 feet tall, with broad shoulders and longer-than-a-man's arms by its side. It had dark hair all over it.

I stopped my car about 30 feet from it, and got out for a closer look at it. I left the door open and the engine running. As I walked toward it, it just stared at me. When I got within 15 feet of it, it turned and walked swiftly into the brush and disappeared. (186)

In the following year (1980), another Indiana resident didn't have to leave his backyard to experience a close encounter with a Bigfoot near Evansville (Vanderburgh County):

I was going to feed the rabbits I raised when I heard someone walking through the brush. When I looked, I saw a dark figure of what looked like a man. I said "hello."

It swayed from side to side and studied me, then began to approach me. I then could see it was covered in black hair.

I turned and ran to the house. I was 20 feet or less from the creature when I saw it. (187)

The next year (1981), a man fishing at night in Knox County encountered a Bigfoot standing in water. The incident began when the witness got an "eerie feeling" that someone was watching him, and looked up to see a pair of glowing red eyes that he assumed were reflecting light from his lantern or a nearby campfire.

The light was insufficient to enable him to see the rest of the face, but the man could make out the top part of a large, hairy body standing in water about four feet deep. The figure was covered in matted brown hair and well built, with unusually long arms.

The witness said the Bigfoot stared at him, tilting its head from side to side. Then it turned, grabbed hold of a tree limb and pulled itself out of the water. As he watched it walk away, he noted that its arms hung down to its knees.

The only sound made by the creature came as it departed. The witness described it as a "loud squeal or high-pitched shriek, something like a young pig would make when you try to hold on to it." *(188)*

Later that year, a Bigfoot paid a scary daytime visit to a rural family in the same county. A mother spotted an 8-foot-tall creature covered with dirty white hair standing in a field about 50 feet away. When the thing started walking toward the house, she hustled her three small children inside. The Bigfoot retreated when the family's dog started barking and ran toward it.

The woman later estimated the creature's weight at about 500 pounds, and said it made a growling sound. She added:

It had a pinkish face and big, glassy eyes. The thing had an awful, sour smell something like dead meat that had set out for three or four days. (189)

Another detailed sighting by an Indiana motorist occurred in 1995 near Corydon (Harrison County), as described by the witness:

I saw something moving, and immediately hit the brakes to slow down. There was an animal crossing the road. It was at least 7 feet tall and probably closer to 8 feet. It was covered with hair that appeared to be about six-to-eight inches long. The hair was a light golden reddish brown.

As the animal walked across the road, it turned to face me. Its eyes were rather large, with big, dark brown or black circles beneath them. Its forehead was slightly sloped. The nose appeared to be slightly flattened.

As the animal walked, it swung its arms like a human. Its arms appeared to be slightly longer in relation to its body than human arms. As it raised its arms, I caught a quick glimpse of its hands. They were very dark in color. The back of the hands were covered with hair, but the palms were clear. (190)

In 1998, a camper had a frightening nighttime encounter with a Bigfoot in the Hoosier National Forest:

I suddenly began hearing leaves and brush rustling. I cautiously approached the ravine where the noises were coming from. I yelled out in the direction of the sounds, and aimed my flashlight.

To my astonishment, (I) viewed a very tall, upright human-looking being hiding behind a clump of trees. The creature stood about 7 feet tall. He was leaning forward and exposing his head to me. He or it was approximately 40 feet from my position.

The light from my flashlight shined into its eyes. I could see that he had dark hair covering his conical-shaped head and dark eyes in the front of his skull. I knew immediately that this couldn't be a man.

I was very frightened, and yelled out again in hopes of scaring him off. He ran away, ducking behind another tree. I continued to follow him with my flashlight and screaming at him. (191)

Chapter 6
BIGFOOT IN THE PLAINS STATES

Iowa/Kansas/Minnesota/Missouri/Nebraska/
North Dakota/South Dakota

Some of the earliest recorded encounters in the Plains States with a creature that sounds like it may have been Bigfoot occurred in **Kansas** in 1869. Here's how a local newspaper reported the incidents at the time:

We of the Arcadia Valley in the southern part of Crawford County are having a new sensation which may lead to some new disclosures in nature history. It is nothing less than the discovery of a wild man or gorilla or "what is it."

It has at different times been seen by almost every inhabitant of the valley, and it occasionally has been seen in the adjoining counties of Missouri. But it seems to make its home in this vicinity.

Several times, it has approached the cabins of settlers, much to the terror of the women and children, especially if the men happen to be absent working in the fields. In one instance, it approached the house of one of our old citizens, but was driven away with clubs by one of the men.

It has so near a resemblance to the human form that the men are unwilling to shoot it. It is difficult to give a description of this wild man or animal. It has a stooping gait (and) *very long arms, with immense hands or claw.* (It) *generally walks on its hind legs, but sometimes on all fours. The beast is as cowardly as it is ugly, and it is next to impossible to get near enough to obtain a good view of it. (192)*

A woman who grew up on a farm near the town of Scammon in Cherokee County (which adjoins Crawford County) recalled an incident which occurred around 1906:

I was about eight years old at this time, and I had a very unusual experience which I will never forget. I was visiting a neighbor girl one beautiful spring day. We were playing along a small stream a mile from her father's home. Across the creek there was an incline.

Suddenly, something came traveling down the incline in a sort of rolling run. It splashed across the creek and jumped into a big clump of witch hazel bushes in front of us. We quickly climbed the fence like a couple of cats. We ran for half a mile and slowed down.

"What did you see?" I asked her. She said, "A big, hairy man with no clothes on." We both agreed he was 7 feet tall and covered with reddish-brown hair. I had nightmares for months afterwards. (193)

More recently, a motorist in the Jayhawk State bravely left her car for a better look at a Bigfoot she encountered while driving near the town of Burlington (Coffey County) one night in 1974.

She brought her vehicle to a stop about 15 feet from the creature after it walked out of a ditch and onto the road in front of her. The woman then directed her high beams at it and walked to the front of car. She observed the creature for several minutes before it resumed walking across the road, stepped over a barbed wire fence and crossed a field into some trees.

She later described the Bigfoot as being between 6 and 7 feet tall and completely covered with hair, although the hair on its face was sparser. It had long arms and glowing red eyes. *(194)*

In 1980, two boys encountered a Bigfoot while riding their motorbikes near Milford Lake in Clay County, as described by one of them:

We topped the crest of a hill and stopped dead in our tracks. At the bottom of the hill, next to a large tree, was someone or something standing there looking up at us. It had a very large, broad body, no neck, large head and was very dark. It could not have been human.

Both of us turned and looked at each other wide-eyed. When we turned back around, it was gone. I don't know if it went back into the trees or into the field.

Later, we decided to see if we could estimate the size of the creature by having one of us stand by the tree. I estimated what we saw was a little over 7 feet tall. (195).

Minnesota was a hotbed of Bigfoot encounters during the 1960s. Like the following incident described by a woman camping with her husband in the Superior-Quantico National Forest on the Canadian border:

I was preparing our lunch and my husband was fishing a few yards down the beach. I heard a noise in the brush and turned around, expecting my husband. Let me tell you, I almost died of a heart attack.

(The Bigfoot) growled. I screamed. He made a gurgling sound. I screamed again. By then, my husband was running up the beach. He arrived in time to see the thing make a rather thorough inspection of our campsite.

He smelled, tasted and then gobbled down a hunk of cheese. He nibbled at a cracker and then put the remainder of the box under his arm. He jerked a string of fish off the holder one by one, and walked back into the woods.

I was struck by his hands. They were not like an animal's paws. These were more human in appearance. (196)

Then there's the Bigfoot that more than qualified a group of Boy Scouts for their wildlife merit badge. Here's how their scoutmaster described the incident, which occurred during a canoe trip on Mantrap Lake:

The thing came out of the woods and walked to the edge of the lake. It glanced up at us, and the boys became frightened that it might charge us.

It was a monster! It had to be fully 7 feet tall and broader at the shoulders than any pro football player. It was covered with long, black hair.

At first, I thought that it must be some massive gorilla that had escaped from a zoo. The gorilla theory just doesn't work, however. This thing had buttocks, and its arms were in better proportion to its body than a gorilla's—that is, its arms were not so long.

After the initial shock of seeing the incredible beast emerge from the forest, the boys seemed to relax a bit and to take in the wonder of it all. Everyone had stopped paddling, as if the sound of the paddle striking water might frighten the beast back into the woods.

The creature stooped to suck up some water, and it drank like a workhorse on a hot day. Every little while it would stop to glance up at us, then it would move its head back down to the lake's surface. When it had finished drinking, it just looked at us stoically.

Its features, although largely covered with hair, seemed definitely human. Yet I am certain that it was not a man. (197)

The Bigfoot which visited a fisherman's shack in Bemidji (Beltrami County) one night may have been looking for some solid food:

I woke up to hear this godawful ruckus in my minnow tank. My dogs were howling and barking.

I snapped on my porch light. That's when I saw this big, black, hairy monster tear away from the minnow tank and head for the woods.

I sure am glad I didn't get in its way. It would have plowed me under. (198)

Another Minnesota sportsman reported an equally upsetting pre-dawn encounter with a Bigfoot at his duckblind near LaCrescent (Houston County) in 1968:

My friend had gone back to the car to get his pocket warmer. I was surprised to hear him come clomping back so soon. It sounded like he was trying to stomp in his footprints so hard the grass would never again grow where he walked.

I poked my head out of the blind and yelled: "Quiet down or you'll scare the ducks back to Canada!"

Well, sir, I found myself looking right in the face of something black and hairy and big enough to take on the Green Bay Packers' defense. It had been looking for roots or something. When it stood up, it just didn't want to stop. I stand 6'3", and this thing left me staring at about the middle of its chest.

I don't really remember pulling the trigger of my shotgun, but I do know that I was not aiming at the creature. It just looked too man-like. It let out a scream and took off for the trees. (199)

The following year (1969), a Minnesota motorist stopped his car on a road near Rochester (Olmsted County) one night for a closer look at the scene of his Bigfoot encounter—and soon had second thoughts about the wisdom of his action. The creature had run into nearby trees at the car's approach. The witness tells what happened next:

I got out of the car and saw that it had been crouched over a dead rabbit. I picked up the carcass.

Suddenly, the calm was shattered by an ear-piercing roar emanating from the woods where the Bigfoot had gone:

The creature must have thought I was stealing its dinner. I wasn't going to argue about it. I leaped into my car, and didn't stop until I got to a police station. (200)

More recently, another motorist was amazed to see a Bigfoot pacing his car while he was driving near the town of Gilbert (St. Louis County) in 1991:

I saw a human-like creature that was at least 8 or 9 feet tall. It had long, shaggy hair. It was running parallel (to the) *side of the car, then it veered off into the woods.*

I stopped the car to see if I could see it again. There was a very bad stench in the air. It was a smell that I couldn't even describe. (201)

One of the more unusual Bigfoot color combinations was described by witnesses in two 1959 sightings near the Nebraska town of Ravenna (Buffalo County). In the first incident, a teenage boy out hunting reported encountering a large, man-like creature covered all over with white fur except for its face and hands, which were black. Later that evening, a carload of high school students said they observed a similar creature in the same area. *(202)*

In 1979, a hunter in the Cornhusker State encountered a creature unlike anything he'd ever seen before near Santee in Knox County:

(It was) *approximately 40 yards away, standing on the edge of a ravine. (Its) height (was) somewhere around 7 feet. (It had) gray hair over (its) entire body, (and a) muscular, very thick neck.*

(It) *was standing very erect, turning (its) head side to side as if trying to smell for something. (It) looked directly at me, and I ducked down. When I looked again, nothing was there.*

A friend of mine was approximately 1-1/4 mile away in the direction in which I thought it went. He told me that I wouldn't believe what ran by his deer stand. He described to me the same thing I saw. (203)

The area where the Nebraska hunters spotted their hairy monster lies just across the Missouri River from **South Dakota**, which saw a flurry of Bigfoot encounters of its own in the 1970s.

In 1974, a man driving near the small Union County community of Jefferson stopped his truck for a better look at a strange creature he spotted walking in the adjoining field. From a distance of about 100 feet, he observed an erect, man-like being that was about 9 feet tall and covered from head to foot with sandy-colored hair.

The witness said the creature was dragging a reddish-colored, furry object that could've been a fox. When the thing spotted the man, it dropped the object, stopped and stared at him. The man got a good look at the creature's face, and was certain it was not a bear. *(204)*

In the late 1970s and early 1980s, there was a veritable Bigfoot invasion of the Standing Rock Indian reservation, which straddles the border between the states of North and South Dakota. Altogether, there were more than two dozen sightings reported.

One witness encountered a Bigfoot at the uncomfortably close distance of 10 feet when he investigated noises in some bushes near his home. The man was shocked to see a hairy creature about 9 feet tall and 600 pounds stand up and look at him. Said the witness:

I didn't stay around long enough to find out what it looked like. I turned around and ran for home, and didn't look back. (205)

In 1994, Bigfoot was encountered in the adjacent Cheyenne River reservation, as reported by a local newspaper at the time:

A group of youngsters spotted a face looking out of some brush. They ran over to investigate, startling four of the creatures, two adults and two young ones. The boys chased them up the (Moreau) river before they headed north.

The adult creatures reportedly were dark brown with a lighter-colored fluffy fringe around the face and a white or gray patch on the stomach. They had human or ape-like faces with oval eyes. The little ones were about the size of a 10-year-old child and the adults were very tall, estimated at 8 foot. They had big arms and were furry all over.

The boys reported that one of the creatures had been dragging some branches and dropped them in the river. There were footprints wider and longer than a human print, with no arch and six widespread toes. (206)

Reported Bigfoot encounters in neighboring **North Dakota** have been fewer in number, but that doesn't diminish the dramatic impact for the people involved.

In 1962, two hunters reported they were the ones being stalked near Minot (Ward County), as related by one of them:

I had the distinct feeling that someone or something was following us. Eventually, I became so uncomfortable that I look(ed) over my shoulder and saw it at a distance of about 50 yards or so.

I thought at first (it) was a great ape. (It had) long, black, shiny hair. (It was) very tall. (It had) no hair on the face above its nose or (on) the palms of its hands. (It had) long arms.

It began to move in our direction. As all I had was a single shot .22, shooting it was out of the question. Both of us ran as fast as we could. We didn't look back after this. (207)

In 1977, a rancher was looking for a stray bull in the area of the Cannonball River near the North Dakota portion of the Standing Rock reservation when he came upon an entirely different kind of animal. He described at as "something 8 or 9 feet tall like a big monkey."

The witness and his son pursued the Bigfoot in their pickup, but without success:

It moved just as fast as a horse. It jumped across a creek and went into the brush, and we lost him. (208)

Bigfoot encounters in the state of **Iowa** hit a peak during the 1970s, dropping off considerably since then.

One evening in 1971, three young men spotted a strange silhouette on a hilltop in a park in Plymouth County near Sioux City. It looked like a large apeman. On an impulse, one of the trio gave his best impression of a Tarzan yell. If his intention was to get the figure's attention, he succeeded admirably.

The thing suddenly started coming down the hill toward them. As they sprinted for their car, they could hear it crashing through the woods. Casting a glance back as they drove off, they spotted it emerge from the woods on two feet.

Later, a gas station attendant near the park reported a carload of shaken teenagers had stopped the same evening and asked for a rag to clean blood off the front of their car. They told him they'd just hit a large, man-like creature which had then gotten up and run off on two legs. *(209)*

Three years later (in 1974), another Bigfoot took a licking and kept on ticking in another nighttime incident near Sioux City. This time, a man who'd heard an odd shrill scream went outside his house to investigate armed with a 30-30 rifle.

He saw a large, hairy creature about 7 feet tall standing 75 feet away. When the thing started walking toward him, the man raised his weapon and fired at it. The Bigfoot toppled over backwards, but then got back on its feet quickly and ran off. *(210)*

In 1977, another Bigfoot was a little too quick for another Iowa man who tried to shoot at it. Prior to the incident, the man claimed to have lured a group of the creatures to his backyard on a number of occasions by leaving food out for them. He said he'd attracted one creature which was 8 feet tall and several others who were about 4 feet tall.

One night, the man decided to shoot one of the creatures to prove their existence. But according to a friend who claimed to have witnessed the incident, the Bigfoot grabbed the man before he could shoot it. The man then blacked out. Before it left, the Bigfoot smashed the windshield of the man's car. *(211)*

About a year after this incident, there were several reported Bigfoot encounters in the vicinity of the Raccoon River in Dallas County. In one, a man approached his cabin near the river and apparently startled a napping Bigfoot, which quickly got to its feet and ran off. He later described the trespasser as about 7 feet tall and covered with black hair. *(212)*

A short time after this incident, a local farmer was returning from his fields one evening when he spied a pair of Bigfoot walking along the road about 100 yards away. One of the creatures was about 7 feet tall, and the other about 6 feet. Both were covered with black hair from head to toe, and had long arms and short necks. *(213)*

Also during the 1970s, two couples reported a terrifying encounter with an uninvited guest at a cabin along the Mississippi River near the town of Clinton (Clinton County). The incident began when several of them went to investigate a heavy, pounding noise on the front porch. Said the owner:

We looked toward the front door as a tall, hairy bulk stepped out of the darkness.

Even more alarmingly, the creature then began tugging at the screen door. The owner quickly grabbed his 12-gauge shotgun:

I always keep the gun loaded, and I emptied both barrels at the thing. It must have been a gorilla or something. It stood at least 7 feet tall, and was the last thing you'd ever want in the living room of your cabin.

Taking the hint, the Bigfoot turned and ran off into nearby woods. Observed the owner:

There was a few drops of blood on the porch. I must have hit the creature, but that didn't stop it, because the gun had light bird shot in it. I may have just stung the thing, although it screamed as if it had been hit hard.

We were damn lucky. The following morning, we tracked the creature and found several large, man-like footprints in a marshy area leading into the woods. Those prints were about 15 inches long. You can imagine what that thing could do to a man. (214)

More recently, there were a series of Bigfoot sightings in the farm country around the small Humboldt County community of Ottosen in the late 1990s. In one, a man observed a hairy, man-like creature outside his home for several minutes after he heard a deep whining sound unlike anything made by the family's farm animals.

He watched as it walked into the barn about 35 feet away, looked inside several spreaders and walked around a grain silo before walking off. The man believed the creature was searching for food.

The witness later described his strange visitor as standing between 6 and 7 feet tall, and said it was covered with black hair. *(215)*

The witness in the incident just described never felt threatened by the Bigfoot he saw. Such was not the case with a woman on another area farm visited by one of the creatures around the same time.

She was washing dishes when she happened to glance out the kitchen window —and found her glance returned by a Bigfoot standing just a few feet away. She said the creature's hairy head had a big nose and "burning red eyes." When she screamed, it lit out running. *(216)*

Bigfoot encounters in neighboring **Missouri** date back as far as the late 1940s, when there were a number of reports of a creature that looked "something like a gorilla" attacking livestock in the southeastern part of the state. *(217)*

In the 1950s, a childhood game in a rural area of Washington County resulted in a memorable experience lasting a lifetime for the witness:

I was 10 or 12 years old. We were visiting my grandmother's house. A bunch of the grandchildren were playing hide-and-seek.

While looking through the cracks of the corncrib to see if anyone had seen me enter, I heard a heavy breathing behind me. Turning around, I saw this huge, man-like creature. (It) was standing over me, just staring at me. He was hairy and huge. It made no sounds, and smelled somewhat like a hog pen.

After wetting my pants, (I ran) like a bat out of hell back to the house. I will swear to this day what I saw was a Bigfoot. (218)

Another indelible childhood memory involving Bigfoot grew out of the experience of a young boy nearly kidnapped by one from his home in Kinloch (St. Louis County) in 1968. He was playing in the backyard when a "gorilla" snatched him up and started to walk off with him. Fortunately for the boy, his aunt's screams and the barking of the family dog caused the creature to drop him. *(219)*

The following year (1969), it was fish-napping that was on the mind of a Bigfoot in Cass County, as recounted by the angler victim:

When I rounded the bend, there was something large holding my stringers of fish. At first, I thought I was seeing things, and closed my eyes and looked again.

There stood what looked like a giant ape or hairy man. It looked like it was 10 feet tall. The arms seemed long, like they could reach its knees. The face was kind of like that of a gorilla, but the mouth didn't seem protruding like that of a gorilla. It looked like its head was set right on its shoulders.

I was scared and froze. It turned and looked at me, and then took off. It jumped the creek and ran up the bank with my fish. (220)

A couple of years later (in 1971), another hungry Bigfoot persuaded two young women to share part of their picnic with it near the town of Louisiana (Pike County), as recounted by the witnesses:

We were eating lunch, when we both wrinkled up our noses at the same time. I never smelled anything as bad in all my life.

I turned around, and this thing was standing there in the thicket. It was staring down at us. It was half-ape and half-man. It had hair over the body as if it was an ape. Yet the face was definitely human.

Then it made a little gurgling sound, like someone trying to whistle underwater. We started scrambling out of there when it stepped forward out of the brush and started toward us. We jumped into the Volkswagen. You can't imagine how fast we locked the doors.

It walked upright on two feet, and its arms dangled way down. The arms were partially covered with hair, but the hands and palms were hairless. We had plenty of time to see this, because the thing came right over to the car and inspected it.

It appeared to have some intelligence, because it seemed to know there was an entrance into the car. This apeman actually tried to figure out how to open the doors.

We were jumping up and down. The car keys were in my purse, which I'd left on the ground when we ran. Finally, my arm hit the horn ring, and the thing jumped straight up in the air and moved back. It stayed at a safe distance, then seemed to realize that the noise was not dangerous.

It stopped where we had been eating, picked up my peanut butter sandwich, smelled it, then devoured it in one gulp. It started to pick up Joan's purse, dropped it, and then disappeared back into the woods.

That was the signal for the women to retrieve the purse with the car keys, and see how fast the Volkswagen could go. *(221)*

The following year (1972), a young man had a scary nighttime encounter with a Bigfoot near the border between Pike and Ralls counties, not far from the scene of the two picnickers' encounter. And once again, the creature's odor preceded its appearance:

I smelled something awful—a cross between a skunk and rotten flesh, like something died.

As I was walking along the side of the road, I noticed someone walking about 300-400 feet away along the edge of the woods. I thought it was kind of late for a farmer to be bringing his cows in. (The cows) were in a big hurry to get away from him.

About that time, I heard a godawful, bloodcurdling scream. I stopped to look at the farmer, and he also stopped and stood still, looking at me. As I continued, he matched my every move, and would stop and stare when I did.

The cows were gone by now, and the stench was getting stronger. What I thought was a man angled his way closer to me. I stopped and stared right at the thing. At that point, I knew it was not a man. It was coming straight for me.

I took off running as fast as I could. As I looked over my shoulder, it stepped over a four-foot-high barbed wire fence without touching the wire. I knew I was in trouble.

I ran for (a) farm house as fast as I could go. I was met by a pack of farm dogs at the gate. I didn't dare look back. The dogs were attacking the monster. (222)

A decade later (in 1981 or 1982), two men made detailed observations regarding the size of the Bigfoot they saw on some railroad tracks in Cass County. They spotted the creature from their parked car, and when they got out for a better view, one of the men inadvertently hit the horn:

The noise caused the Bigfoot to spin around. Sighting us, he dropped flat to the ground. It filled the rails with its shoulders. If I were to lay flat on the ground, I would be about 12-13 inches thick from my back to my chest. This thing was easily three times that.

We watched it for a few seconds, when it stood up. It took a step towards us. That is all it took for us to leave quickly.

I would say that it was between 7 and 8 feet tall, and possibly 400 pounds. It was dark brown and shaggy-haired. The head seemed to rest on the shoulders, or there was a very broad and well-muscled neck holding it up. (223)

One night in the late 1980s (either 1988 or 1989), two sets of startled motorists encountered a Bigfoot roadblock on a highway near Waynesville (Pulaski County), as described by one of the witnesses:

We rounded a corner and were meeting a car, and there in the middle of the road is this "thing." It was about 7-8 feet tall, and covered with hair. It was "frozen," much like a deer is when headlights hit it.

My husband slammed on the breaks, as did the car that we met. We were just past the creature, as was the other car. There standing on the yellow line, illuminated by the taillights of both vehicles, stood this creature. It stood there for about 30 seconds, and then sort of leaped off the roadway. (224)

A couple of years later (in either 1990 or 1991), two people had an unexpected visitor to their campsite on a farm near the small Pike County community of Cyrene just before dusk, as described by one of them:

There was an awful smell. I looked and saw a creature that looked like a large ape or bear. It looked right into my eyes.

I couldn't move. We stared at each other for what seemed like forever, but it was probably about 30 seconds.

Then it turned its head, and from a squatting position, it jumped up an 8-foot bank and ran in the other direction. (225)

In 2000, a Missouri motorist had a close encounter with a Bigfoot on a road in Ste. Genevieve County, as related by his daughter:

(It) *crossed the road just in front of his vehicle in broad daylight. He described it as very tall, but hunched over.* (It had) *different shades of hair all over its body, with very long arms and an ape-like face. He also said* (its) *stride was not human-like, but sort of a limping stride. (226)*

Chapter 7
BIGFOOT IN THE SOUTHWEST

Arizona/Arkansas/Louisiana/New Mexico/Oklahoma/Texas

Arkansas can lay claim to the longest-running Bigfoot tradition in the Southwest, with reports dating back nearly 150 years.

The following account is from a an 1851 issue of a New Orleans newspaper:

Mr. Hamilton of Greene County, Arkansas, while out hunting with an acquaintance, observed a drove of cattle in a state of apparent alarm, evidently pursued by some dreaded enemy.

They soon discovered as the animals fled by them that they were followed by an animal bearing an unmistakable likeness of humanity.

He was of gigantic stature, the body being covered with hair and the head with long locks that fairly enveloped the neck and shoulders.

The "Wild Man," after looking at them deliberately for a short time, turned and ran away with great speed, leaping 12 to 14 feet at a time. His footprints measured 13 inches each. (227)

In the early 1920s, a group of people were picking berries near Booneville (Logan County), when they were startled by the appearance of a Bigfoot. The following details were gathered by the grandson of one of the witnesses:

(A) *hairy, man-like creature was after my great-grandfather's horse.* (The) *horse was trying to pull away from the tree he was tied to. My great-grandfather heard the commotion* (and) *approached his horse.* (He) *observed this creature, which then ran into the brush.*

He told me this "thing" was taller and heavier than he was. My great-grandfather was 6'2" and 190 pounds. (228)

Spooked horses were also a feature of a Bigfoot encounter which occurred in 1966 in an area near the War Eagle River in the northwest corner of the Razorback State. The incident began when two men riding horses met a farmer coming the opposite direction who appeared to be driving his tractor as fast as it would go. He stopped only long enough to warn the horsemen that he'd just seen a horrible monster in the field ahead.

The two men decided to see for themselves, but their mounts obviously wanted no part in the venture. The horses became agitated and refused to go on, so the men tied them to a tree and continued on foot.

They soon reached the field, where they spotted a large mound of white fur on the ground. When they approached to within about 10 feet of it, a 9-foot-tall Bigfoot suddenly got to its feet. It was completely covered in white hair about 3 inches long, except for its face and the palms of its hands, which were pink and hairless. The witnesses said the creature gave off a strong, offensive odor.

The men's observations were cut short when the Bigfoot started walking toward them, emitting a strange beeping sound. Their curiosity now apparently satisfied, the pair quickly retreated to their waiting horses and rode from the area. *(229)*

The area around the town of Fouke (Miller County) has been the scene of a number of Bigfoot incidents over the years, even inspiring a low-budget horror film (*The Legend of Boggy Creek*).

One of the most dramatic Bigfoot encounters in the Fouke area occurred in 1971, when a woman sleeping on a sofa in her living room was awakened by strange noises:

(I) *saw the curtain moving on the front window, and saw a hand sticking through the window. At first, I thought it was a bear's paw, but it didn't look like that. I could see its eyes. They looked like coals of fire—real red.*

The woman screamed at this point, bringing her husband rushing into the room. They both saw a creature they described as "about 6 feet tall, black and hairy" running off on two feet.

Later that night, the husband decided to look around outside their house—a decision he soon had cause to regret when the creature suddenly came up behind him in the darkness and grabbed him. Somehow, the terrified man managed to break free. He was so intent on reaching the safety of his house that he dashed through the front door without taking time to open it first. *(230)*

In 1988, a motorist reported a nighttime encounter with Bigfoot on a road near Jonesboro (Craighead County):

I saw a creature about 7 feet tall, weighing probably 300 pounds and having thick fur. The creature was running down the roadside ditch, and then it came out of the ditch and ran upright across the highway and disappeared into a patch of woods on the other side. It took only three strides to cross the highway.

I don't believe it was a bear or a human, because bears usually run on all fours, and this creature ran upright with its knees slightly bent, unlike humans. I saw something that night that I have never seen before. (231)

In 1994, a homeowner near Benton (Saline County) who thought he was chasing deer from his property was surprised to find himself confronted by a much more formidable form of wildlife:

The deer were always coming up to our back fence and causing these little wimp dogs of mine to go into cardiac arrest, especially at night when I was trying to go to sleep. So I was in the habit of taking a small rifle and shooting a round into the trees to frighten the deer away and shut the dogs up.

One night, I decided to (go) into the woods to really scare them away. About 30 yards into them, I heard movement. Shining my light, I saw the thing standing rock still.

I could tell it was about 7 feet, and it was half again as broad across the shoulders as me. It had brownish-orange, long fur over what I could see of its body, and its face was dark.

It grunted and made a move towards me. Being a redneck, the first thing I did was aim and fire. I know I didn't miss, because he was only about 30 to 35 feet away, and I saw fur fly.

It yelled and began to move towards me again. I began to move backwards, reloading the gun, and never taking the light off of it. I fired again, not missing, and he really started to move on me. I turned and ran to my house.

I could hear it running through the woods. It let out a loud yell that reminded me of sounds you might hear at the primate house at the zoo. (232)

Another southwestern state with a long and colorful history of Bigfoot encounters is **Oklahoma**.

About 1915, a young man was returning to his home one night near Wann (Nowata County) when he encountered a strange creature standing by the gate. Like other witnesses unfamiliar with Bigfoot, he had difficulty reconciling what he was seeing:

It was about 5 or 6 feet tall, and it stood with its arms stretched out. It was about four feet wide in the chest, and hairy all over. It was like a bear or something, but it stood up like a man. (233)

A woman who was living near Gould (Harmon County) in 1942, provided this account of a Bigfoot encountered that year by two members of her family:

My mother and sister were going to get the cows from the pasture to bring them to the barn. They had just gone through a shelter belt of trees into a clearing, when they heard rustling like something moving in the trees.

Thinking it was one of the cows, they turned around and saw a very tall, hairy creature running away from them. It was standing completely upright, running on two legs. Needless to say, they ran the other direction. (234)

Four years later (in 1946), Bigfoot was blamed for a deadly attack on a man in Jefferson County. The victim and a female companion were making out near their parked car in a remote area near some woods, when their lovemaking was suddenly interrupted by loud screams and the sound of something crashing through the nearby woods toward them.

The pair wasted no time climbing into the car and driving off. Assuming they were safe, the man obeyed a stop sign about ¼ mile down the road. But before he could drive on, a huge hairy arm smashed through the side window and grabbed him by the throat. His panicked companion stomped on the accelerator, and the car roared off. Unfortunately, the man sustained a fatal injury to his neck in the process. Local police classified the man's death as an "unknown homicide." *(235)*

In contrast to this incident, the Bigfoot seen by a youngster outside a home in Jenks (Tulsa County) in 1952 seemed merely curious. It nevertheless proved a traumatic experience for the witness:

I noticed a large (maybe 7 feet tall) creature covered with brown hair looking in the window. He had his forearms resting on the windowsill, and was staring at grandma. I jerked the covers over my head and played dead.

I will never forget the fear that gripped me that night. I knew I had to get up and go get help for grandma's sake, so I threw off the covers and started to run into the living room and tell dad.

I noticed that the creature had turned and was walking away toward the creek. He was very broad across the back, extremely muscular and covered with slick, brown hair. (236)

Another Bigfoot may have turned aggressive because two young men invaded its territory near Broken Bow (McCurtain County) in 1971:

We became aware we weren't alone. Just beyond our vision, we could hear something walking. When we walked, it walked. When we stopped, it stopped. The worse odor either of us had ever smelled was nearly overpowering.

Suddenly, we heard it moving toward us. We began to run. I'm not ashamed to say I was scared.

We found ourselves standing on the edge of a drop off of about 10 feet, and whatever was pursuing us was coming closer. I screamed to jump. We landed on our feet and tried to catch our breath.

Then we heard it—a grunting growl (and) *heavy breathing. We both turned our heads upwards and looked into the face of what I at first thought was a gorilla.*

We ran through thorn bushes. We crashed through trees and brush a sane man would have gone around. All the time we could hear this "gorilla" crashing through the same brush and trees. Only it sounded like a freight train was coming after us.

To this day, I don't know how we made it to the beach, but we made it. However, our escort was still coming, and it didn't sound too pleased. Fortune smiled on us. We came out only yards away from our boat.

As I put the boat in reverse and was moving away from the island, "it" burst out of the brush and onto the beach. It made a scream I will never forget. It entered the lake after us, swinging its long arms (and) *splashing water.*

It was about 8-9 feet tall (and) *heavily built.* (It was) *covered in reddish-brown hair which covered its entire body except around the eyes, nose and mouth. I had the impression its head was conical in shape. I can't remember seeing much of a neck. (237)*

Beginning in 1974, Bigfoot became something of a regular visitor for a number of years to the area around the Nowata County communities of Noxie and Nowata.

One woman who had multiple sightings during this time—once from as close as six feet and another time for a period of several hours—provided the following detailed description of a 7-foot-tall Bigfoot she saw:

(The) *face* (was) *devoid of hair except where a man would have a beard. He seemed to have no neck. His head rested on massive shoulders, from which long, powerful arms hung nearly to his knees.*

He was a magnificently-built creature with a wide, deep chest, narrow hips and thick, muscular legs from which grew 18-inch-long, eight-inch-wide, hair-covered feet complete with toenails. (238)

Some of the woman's neighbors had more frightening encounters with Bigfoot. Like the two men who went to investigate scratching noises coming from an abandoned building near the home of one of them in 1975:

We walked over by that house, and it was standing there watching us. We walked towards it, and it started growling at us.

I'd say it was 7 or 8 feet tall. It had hair all over its body a dark blackish-brown color, anywhere from an inch-and-a-half to two inches long. It had hair all over everything but around its eyes and its nose. That was the only part where I could see the skin on it.

The eyes glowed in the dark—reddish-pink eyes. You don't need a light or anything to shine on them like most other animals. They glow without a light being on them.

We was about 10 foot away from it. We just ran away, and it ran, too. (239)

Another area man may have owed his memorable 1983 encounter with Bigfoot to a charred dinner he discarded near his campsite one night. He awoke to the sound of chicken bones breaking, and spotted a 7-foot-tall figure in the light of his campfire:

I picked up my .22 and popped off a shot over its head. I didn't want to shoot it. I wasn't sure if it would come towards me or what.

The man turned his back to the creature to gather up some of his gear, when he was struck on the back of the head:

Whatever it was, it felt hard. It was like somebody doubled up their fist and hit you. It took me down to my knees.

Fortunately, the witness was able to jump to his feet, sprint to his truck and drive off. *(240)*

That same year, a family living near the town of Wilson (Carter County) had a harrowing Bigfoot experience after the man took a shot at one.

According to the man, the Bigfoot started the fracas by walking around the outside of the house pounding on the walls and breaking windows. The man retaliated by shooting the creature through a window with a 12-gauge shotgun, causing the Bigfoot let out a loud yell.

But if the man thought that would make the creature go away, he was badly mistaken.

The shot only served to enrage the Bigfoot, which then tried to break down the back door. Wisely, the man hustled his family out the front door and into his pickup truck.

The Bigfoot apparently succeeded in gaining entry after the family drove off, because when officers later returned to the house to investigate the family's story, they found a scene of such devastation that it almost looked like a tornado had struck. Among the items damaged was the shotgun which the man had used to fire at the Bigfoot. Its barrel was now twisted like a pretzel. *(241)*

In 1999, a man claimed he was attacked by a family of Bigfoot creatures while walking in a wooded area in the southeastern part of the Sooner State:

I got a very uneasy feeling. As I turned to head back out, I was struck on my shoulder and knocked against a tree. Three (Bigfoot) were 20 feet away from me and leaving the area. On the ground was a rock the size of a softball. I am not sure if they hit me with this rock or struck me with their hand.

I observed as much as I could in the short time I had to view them. One was a large male from 10-12 feet tall. He had a crest on his head, was a dark brown color and had gray in his hair. The crest was a lighter color like auburn on top and turned more gray as it went down his back. You could see dark, brownish-black skin under the hair.

The hair was long on the head, and hung onto the shoulders. The tops of the shoulders had long hair growing off of them as well, like long-haired shoulder pads. The rest of the hair was approximately four inches long. I found his track later, and it was 19 inches long. He had a flat foot, and took strides of a little over six foot between steps.

The female one was a lot smaller. She was about 6'3" or 6'4" tall, and was a dark reddish color. She had a crest on her head as well. She also had longer hair on her head and shoulders, just like the larger one did. She had no hair on her buttocks, and reminded me of an older, obese woman going up the hill, as her backside was fat and wrinkled-looking. The skin color was a dark brown under her fur.

The third creature, a male, was either a very sick juvenile or a real old creature. It was grayish-brown color. It was colored so that if it stood against a tree, you would not notice it. It was very thin. It was about 6 feet tall and would have done good to have weighed 150 pounds. The hair was missing in patches all over the thing, and the skin was hanging off it as well. It didn't have any hair on its rear end, either, and not a whole lot on the rest of its body. (242)

Later in 1999, three young people encountered an aggressive Bigfoot one night while camping near a lake in Washita County. The incident began when they noticed a pair of red eyes peering at them from out of the darkness, and one of the young men yelled.

The shout brought a swift reaction in the form of a shaggy-haired, 8-foot Bigfoot which came lumbering toward their campfire. The trio wasted little time making a strategic withdrawal to the young woman's nearby truck. Their haste proved well-advised. The creature chased them as they drove off, scratching the vehicle's tailgate and bumper. *(243)*

Texas has reported Bigfoot encounters dating at least as far back as the 1930s.

The following account is from the son of a man who experienced an incident near Red Oak (Ellis County) in either 1938 or 1939:

He and a few friends got together to go coon hunting one night. They brought with them two coon dogs and a .22 handgun. When they made it to a place where they could camp, they started a campfire. The dogs were acting very scared, and were close to the fire.

One of the men noticed a huge, white figure standing about 30 yards away, just watching them. He pointed out the creature to the rest of the men. They just took off running as fast as they could, dogs and all.

My father's bother-in-law got a few of his friends together and drove down to the area. After a couple of hours, they heard something moving around in the brush behind the car.

It was then that this huge, white-haired creature stepped out and started heading toward the car. The driver started the car, and they drove out of there as fast as he could. The creature chased them all the way to the main road, where they finally lost it. (244)

Things were quiet on the Bigfoot front in the Lone Star Sate for the next three decades, but resumed with a bang in 1969. One focal point of Bigfoot activity that year was Lake Worth, located on the edge of the city of Fort Worth and adjacent to a park and wildlife refuge.

In one incident, several carloads of people stopped to observe a 7-foot-tall creature covered with whitish-gray hair and weighing an estimated 300 pounds as it stood on top of a hill. In a reversal of the usual roles played in most encounters between Bigfoot and humans, the people started moving up the hill toward the creature. They were soon discouraged from advancing further, however, when the Bigfoot picked up an automobile tire with rim attached and hurled it some 500 feet in their direction. The creature then ran into the brush and disappeared. *(245)*

A few months after this incident, a man was sleeping in the bed of his pickup truck on the shore of Lake Worth when he had a rude wake-up call at the hands of Bigfoot. The creature—which the man described as "a cross between a man and a gorilla"—reached into the truck, grabbed him sleeping bag and all, and unceremoniously dumped him on the ground at its feet.

Acting with a resourcefulness that would've made Colonel Sanders proud, the man picked up a bag of leftover fried chicken and thrust it at the Bigfoot. Whereupon the creature emitted a few guttural sounds, turned and ran into the woods. The witness claimed he then observed the Bigfoot enter the lake and swim to an island. *(246)*

It was also in 1969 that a Bigfoot threw a big scare into two men one night while they were parked along a levee near Commerce (Hunt County):

He came over that levee squalling and tearing out saplings and ripping up tall grass and heading toward the car.

And it was big. I've seen some people 7 feet tall, but this thing was bigger than any man I've ever seen in my life.

You could have stretched a yardstick across its shoulders and its shoulders would've been wider than that yardstick. It was big and hairy, whatever it was. I'd never been so scared in all my life.

The pair didn't stick around long enough to make any further size comparisons that night, but they returned the next day and found some oversized footprints:

I put my arm down in one of the prints, and that print was as long as from my elbow to the tip of my outstretched fingers. (247)

Two years later (in 1971), two other men encountered a similar creature neither of them would soon forget in adjacent Lamar County near the town of Paris. The pair were driving a butane gas delivery truck one night, when they stopped along a rural road to relieve themselves.

Suddenly, a hulking form emerged from out of the shadows and walked toward the truck. One of the witnesses had just enough time to notice that their hairy visitor's head was level with the top of the truck's butane tank, making it more than 8 feet tall:

That thing wasn't more than four feet from me when I dove into that truck. I'm a hunter, and I'm not scared of the woods or anything in it, but that thing reached out for me and I was afraid for my life. I don't know what I'd have done if it caught me. (248)

In 1976, a pair of Bigfoot creatures struck fear in a man and his horse from a much greater distance one evening near Gladewater (Upshur County):

The horse started to prance, snort and blow. He was watching something off to my left. This horse had a long mane, and I noticed it standing part way out like it was charged with static electricity. The hair on my arms and back of my neck were standing up, too.

There was a full moon. I could see in an opening in a meadow two large creatures standing erect and side by side, watching me on the horse. They appeared to be swaying back and forth like they were trying to scent what was coming down the road. One looked to be about 7-8 feet tall, and the other 6-7 feet tall.

I started to get scared even though they were on the other side of a barbed wire fence and about 35 yards from the horse and me. I kicked the horse into a gallop and headed home, my heart pounding in my chest. I looked back over my shoulder, and saw them disappear slowly, still walking erect, into the woods. (249)

A couple of years later (in 1978), Bigfoot was back in the area of Hunt County around the town of Commerce. In one incident, a driver nearly collided with a 7-1/2-foot-tall specimen on a country road at night:

I slammed on my brakes and came within a few feet of hitting the thing. It turned slightly toward my car. Then, in one step, it was across the road. I hope I never run into anything like that again. (250)

Two nights later, a Bigfoot made a big impression on three high school students who were walking along the same road. The trio stopped when they heard a noise near a tree at the side of the road:

The creature jumped up from behind the tree. We were about three or four feet away from it when it came up, and we just ran. I don't know if it came toward us or went the other way. I was too busy running.

It had to have been well over 7 feet tall. It wasn't a gorilla. (251)

The following year (1979), a family had an uncomfortably close daylight encounter with a Bigfoot near Marshall (Harrison County). A couple and their two young daughters were shooting a target pistol in a ravine when:

Suddenly, a large animal lunged into the ravine only about 10 or 15 feet in front of us. This animal was grayish in color. (It had) hair about 5-6 inches long. It wasn't very thick hair. You could almost see skin through (the) hair.

(It had) very wide shoulders. His upper arms were very long, linear and powerful-looking. His thigh bones were also very long. He had a short neck and round head.

He paused, looked at us, and then in one leap jumped out of the ravine. We froze for a moment. Then (my) husband grabbed (our) six year old, and I took the hand of (our) nine year old. We ran towards the car.

The loud brush noises indicated the animal was running toward our car. It stopped in brush immediately in front of our car. We shoved (the) kids in, climbing in behind. (252)

In 1985, a motorist took a wait and see attitude toward the Bigfoot she encountered on a road near Hamilton (Hamilton County):

A huge thing came out the side of the road and got right in front of the car, and stood up on its hind legs.

I stopped and locked my doors, and waited to see what it was going to do. It looked right at me. It had a face of an ape. It was (a) big, black something. (253)

Another motorist provided a more detailed description of the creature he and his family encountered in the same area one night in 1992:

It stood approximately 7-to-8 feet tall, weighing between 400 and 500 pounds. Its body was covered with hair. It had long arms that extended down to its knees.

Its face looked almost human. It looked at us and growled a low moan, showing four fang-like teeth, two on top (and) two on (the) bottom, and the rest flat like humans'.

Then it hurtled over the guardrail and ran off into the night toward the brush along the riverbank. (254)

In 1995, a man claimed one of the closest encounters ever with a Bigfoot in the Lone Star State in the Sam Houston National Forest (San Jacinto County). The witness was inspecting some property on foot, when he nearly stepped on a strange creature lying on the ground, resting on one elbow. The two stared at each other for a few moments.

Then the creature stood up to a height of about 7-1/2 feet, struck the man on the chest with one of its arms (knocking him down) and ran off on two legs.

The man was able to describe his assailant in some detail. He said it was covered in brownish-black hair everywhere except around the eyes, nose and mouth and on the palms and soles of the feet. It had a sagital crest, and the hair appeared coarse and stiff. It had broad shoulders and no visible neck.

The creature's eyes were gray and human-shaped, but were set farther apart than a man's. The nose was more human than ape. The lips were wider and fuller than a person's, and the teeth were large, square and discolored. The ears were generally human-shaped, but perhaps more pointed.

The creature, which was definitely a male, had long arms and large hands with long, thin fingers. Its feet were basically shaped like a human's, but were 15-to-17 inches long and appeared to have little or no arch in them.

The witness could hear the Bigfoot breathing deeply and making grunting sounds. He also noticed a strong wild animal odor that reminded him somewhat of a bobcat or wolf. *(255)*

In 1998, a couple also managed to note a number of details during a nighttime encounter with a Bigfoot they observed in their headlights while they were driving near Greenville (Hunt County):

I noticed something very large and dark. The creature was walking upright. (The) arm length appeared to be below the knees, although its arms were swinging back and forth as when someone is walking. The creature's posture had the upper part of the body leaning forward while it was walking, instead of straight up like a human. The steps covered what looked to be four-to-five feet per stride. It also looked to be walking at a face pace.

It had hair that was very dark all over its body. The creature had an odd-shaped head. The head appeared to come to a point towards the back of the head. I did not see the face. My wife noticed what appeared to be two eyes which showed to be red.

There were several orange flags like those used by surveying crews hanging in a tree (6 feet) off the ground, and the creature appeared to be at least two feet taller than that. (256)

Many of the Bigfoot encounters in neighboring **Louisiana** have taken place in that state's extensive swamplands.

In 1962, two hunters came upon a strange creature in the Honey Island Swamp, on the border with Mississippi. It was rooting in mud about 30 feet away. When it became aware of the men's presence, the thing stood up on two feet and stared directly at them.

The witnesses described the Bigfoot as having a gray, hair-covered body with a huge chest and shoulders, and man-like facial features. Before the witnesses could note any further details of the creature's appearance, it turned and walked off into the bushes. (257)

Two other veteran hunters got a more detailed look at the Bigfoot they encountered in the same swamp in 1973:

That's when we saw this thing. It was standing with its back to us. My friend and I both stood and stared at it. Neither of us had ever seen anything like it before, and we had trouble believing our eyes. Then the thing turned around and looked at us. It was ugly and sinister. Sort of like something out of a horror movie.

I'm sure it was at least 7 feet tall, and it must have weighed 400 pounds. The hair on its head hung down about two feet. The rest of it was covered with short, dingy gray hair. Its chest and shoulders were massive. Its face was square, and I could see two rows of teeth in its powerful jaws.

The thing must have stood staring at us for a full minute before it went tearing off into the woods. I want you to know it scared me real bad. (258)

In another incident in the Honey Island Swamp around the same time, a small boat carrying a fishing guide and another man apparently collided with a submerged Bigfoot:

I thought we might have hit an alligator or a big turtle. I stopped to take a look in the water. I couldn't see what we hit, but it felt like it was an animal of some kind. Then my client shouted and pointed toward shore.

We both saw this big, gray, furry creature scrambling out of the water. Once it reached the bank, it immediately raced off and disappeared into the swamp. (259)

In a 1975 incident near Oakdale (Allen Parish), a family dog provided early warning of the presence of Bigfoot:

All of a sudden, the dog ran toward me. (It) was scared to death. I knew something was wrong, because the dog was never scared of anything.

I looked up, and I saw a tall, hairy-looking beast. There was a strong odor.

It scared me so bad, I ran to my house. By the time I got to the house to tell my family, the creature was gone. Whatever it was, it didn't want to hurt me or it would have. It had plenty of time to get me. (260)

The following year (1976), an encounter with Bigfoot took some of the charm out of country living for a Louisiana woman residing near the border with Arkansas:

One day at sunset, I threw out some mop water in the backyard. Running across the clearing from one thickly-wooded area to the other was a thin, hunched creature covered with light gray hair.

Very long arms swung gently along its body, and its hands were cupped back as it ran. It was shorter than a Bigfoot should be, unless it was young.

I quickly went into the house and locked the door. We moved back to the city that summer. (261)

In 1980, a hunter encountered a larger Bigfoot near Oakdale (Allen Parish):

I was hunting with my dog when I got a strange feeling I was being watched. My dog ran into a thicket (and) he started barking. Then I heard a scream. It was so loud, I could hear the sound over the dog barking.

The dog quit barking, and I could hear sounds of snapping twigs and limbs. My dog came running out of the brush right past me in the direction of my home. I called him, but he wouldn't stop,

I turned back toward the area where he came from, and what I saw scared the hell out of me. I was standing about 25 yards from the thicket. I could make out the outline of something standing in the brush. It was 7 feet tall or more and very large. It seemed to be looking at me.

It began to move toward me. I could see the brush moving around it, and then I knew it was alive. I turned and ran, never looking back. (262)

Two motorists had a brief but memorable encounter with Bigfoot on a back road between Dulac and Chavin (Terrebonne Parish) one night about 1990, as recalled by one of them:

All of a sudden, this huge, white Bigfoot came across the road no more than 50 feet in front of us. (It) took two big steps across the road. His or her first step was almost in the middle of the road. I mean this thing was big—around 7-to-7-1/2 feet tall and at least 300 pounds.

We stopped the truck, and my cousin grabbed his spotlight and pointed where the thing went. This thing moved fast. (We) saw where bushes and small trees had been pushed right over like nothing. (263)

Three years later (in 1993), a squirrel hunter definitely lacked sufficient firepower to consider tangling with the super-sized Bigfoot he encountered in the northwestern part of the Pelican State. The man heard the sound of some large animal approaching his position one morning, and thought it was probably a deer. What he saw standing next to a tree about 25 yards away was no deer —or any other animal he'd ever seen before.

It was a huge, hairy, figure standing on two legs. The man estimated its height at between 11 and 12 feet, and thought it must have weighed at least 600 pounds. The creature was covered in short, dark-colored hair, and had a large head with dark eyes. The witness could clearly see it was male.

As he watched, the Bigfoot grabbed a large pine tree and shook it so forcefully that pine cones came cascading down. At this point, the man decided he'd seen enough, and retreated from the area. *(264)*

Bigfoot encounters in **Arizona** have been few and far between. Among them is this account of a vegetable raid which occurred near Flagstaff (Coconino County) in 1924:

We were walking along halfway between the house and garden, when my mother looked up and saw what she thought was my husband. I looked, too, but what I saw didn't look like my husband.

"It" had an armload of corn, and was bent over, pulling up turnips. My mother called, ran forward a little distance, then waved.

The "thing" stood up straight, and I figured it was about 7 feet tall and weighed about 400 pounds. It was light in color, and seemed to have a hairy body.

It took off through the wheat field and jumped over the rail fence, disappearing in the thick forest of young pine trees. (265)

Four quail hunters encountered a Bigfoot near the Santa Rita Mountains in the southern part of the state in 1983:

We heard a gunshot, and something came up over the hill in front of us, ran down the hill into the gully, and continued running into a canyon. I looked at my nephew and asked him if he'd seen what I had just seen. We both kind of stood there in amazement at what we had seen.

What we saw was something with medium brown-colored fur approximately two inches long. The creature had long arms, long legs and no apparent neck. It appeared to be about 6-1/2-to-7 feet tall. We (found) footprints in the sand about 14 inches long. (266)

The southwestern state with the youngest Bigfoot tradition is **New Mexico.**

In 1966, residents in suburban Albuquerque (Bernalillo County) lodged a number of complaints with local police concerning incursions onto their property by a hairy intruder which came to be known as the Cry Baby Monster, named for its vocalizations resembling an infant crying.

A motorist nearly hit the thing when it suddenly appeared in the beam of his headlights one night:

I hit the brakes hard. I thought I was going to hit someone. It was about 5 feet tall, looked part human and cried like a baby. It wasn't a man, and it certainly was not a gorilla. It stood there, transfixed in our headlights, and then shuffled off into the nearest yard. (267)

Several residents reported backyard altercations with the creature. In one, a family's cat engaged in a brief howling and spitting match with it before scurrying indoors. In another, a young man was knocked senseless when he went outside to investigate a nocturnal prowler. When he came to, he found unusual footprints on the ground. The incidents continued for about a month and then ceased as abruptly as they'd begun. *(268)*

In 1980, the Land of Enchantment lost some of its enchantedness for a Navajo family living near the San Juan County town of Toadlena. The incident in question was preceded by a series of strange happenings, as described by one family member:

The dogs would bark at something unseen to us, and the sheep would run away from something hidden in the forest. We would feel like something was watching us from the shadows.

Then, one evening around supper time, the shapeless something suddenly assumed a surprising shape. The incident began when another family member burst into the house:

(He) *came running in crying and white as a sheet. He was always the macho type that showed no emotion, so to see him this frightened alarmed* (us) *to hurry out the door.*

I saw something down at the well. I first thought it was a black bear. Then I noticed it was a grayish-brown color and kneeling. I never heard of a bear that could kneel like a man. It looked as if it was washing something off or taking a drink.

I had no problem with it until it looked over its shoulder at us, and I could see its features. (They were) *man-like. (269)*

Chapter 8

BIGFOOT IN THE MOUNTAIN WEST

Colorado/Idaho/Montana/Nevada/ Utah/Wyoming

Most recorded Bigfoot encounters in **Wyoming** have occurred since the 1970s.

In 1972, two teenage boys told sheriff's officers on the Wind River Indian reservation (near Lander) that a Bigfoot chased them while they were riding their horses one day.

The duo's haste to leave their pursuer in the dust precluded a detailed description, but they were able to note that the creature took long, five-foot strides and had one of its hands tucked beneath its arm as if injured. Authorities later found large, man-like footprints in the area where the boys said the incident occurred. *(270)*

In 1978, two motorists got a much better look at the Bigfoot they encountered one night while driving between the town of Cody and Yellowstone National Park:

As we came around a curve in the road, our high beams illuminated a large, dark, shaggy figure coming up out of the ditch on the side of the road at a distance of about 200-250 feet.

As we approached the figure, it looked first at the vehicle (we noticed the yellow reflection from its eyes that is seen in a dog's eye when light catches it at night), then deliberately turned its head away from the lights. That motion was non-human or bear-like, in that the shoulders, chest and head moved simultaneously.

We slammed on the brakes, stunned at what we were seeing—a hominid creature perhaps 7-to-7-1/2 feet tall, massing perhaps 600-to-800 pounds, standing completely upright.

The head appeared to merge into the neck. There was no snout or protrusion from the face as would be commonly seen in a bear. The face was not clearly visible, and was only glimpsed for a moment. We both got an impression of long hair covering some of it.

It was heavy and powerful-looking. It possessed a rather blocky, yet elongated head, slightly domed on top of the cranium, (a) thick, short neck, broad shoulders (and) full chest. It was square and longer through the torso and hips than a human.

As it walked across the road in front of us, the buttocks were clearly seen as muscular masses moving under heavy, shaggy hair, attached to long, powerful, muscular thighs longer in proportion to a human, big knees that functioned as a human knee, thick, muscular calves and feet in proportion to the rest of the oversized body.

The soles of the feet appeared to be hairless or less covered in hair and very dark in color. The arms hung from heavily-muscled shoulders, and were longer than a human, reaching to knee-length. The elbows were perhaps a little further down the arm than on a human.

It took three extraordinarily long and fluid strides across the highway, appearing to catch hold of the metal barrier with one long-fingered, hairy hand and swinging down into the culvert. (271)

A couple of years later (in 1980), two other men on foot claimed they were pursued into the town of Jackson (Teton County) by an inhospitable Bigfoot when they ventured onto a nearby mountain. According to a report in an area newspaper:

Two men reported to police that they were chased off Snow King Mountain by a "Bigfoot-type creature" 12 feet tall with long, dark hair and arms which hung almost to the ground.

The men told police the creature breathed heavily and made a moaning, growl-type noise. They described the creature as having a simian-like face as big as a stop sign, and (said) that the creature was humpbacked.

(They) told police that they ran when they spotted the creature, and that it followed them. The last time they saw the creature, it was standing under a streetlight.

The police reported that the two men had not been drinking. (272)

In 1985, three young men didn't wait to be chased by the Bigfoot they encountered in a wilderness area near the town of Shell (Bighorn County):

I just froze at what I saw. At first, I thought it was a large bear, but discounted that idea immediately. It was a massive, human-like animal covered in dark, coarse-looking hair. The face, chest, inside elbow area and hands were nearly bare.

It was around 8 feet tall, and I would say it weighed between 450-550 pounds. It had shoulders that were extremely wide, and they sort of slumped forward. Its arms were phenomenally long and thick. They hung far below the thigh area. The buttocks were unproportionately large. They, too, were very muscular-looking.

The head of the animal seemed to be plopped right onto the shoulders. If the thing had a neck, it wasn't any more than three-to-four inches long. Its face looked like a Negro, but the browline and nose were more pronounced. The lips seemed thinner than that of a Negro. The face was somewhat like a black man's and somewhat like a gorilla.

We were about 15-20 yards from it. It stood there seemingly observing us, and we crouched observing it. This went on for at least two minutes. Then it made a noise—not really a growl, but more like a deep cough, like it was clearing its throat, but louder.

That did it for us. We turned and sprinted down the fenceline. (273)

In 1997, a family leaving Yellowstone National Park spotted an unexpected example of the local fauna:

As we were driving out of the park, I was looking up the mountain to see if I could spot bighorn sheep. I noticed, walking in large strides, a tall 8 or 9 foot, hairy, upright Bigfoot-like animal. It was so tall that you couldn't help but see it.

Then it made three strides across this rocky terrain and stopped just above a green, grassy-like area next to the snow.

My son saw it the same as I did, because he was excited, saying it looked like and walked like Chewbaca, the "Star Wars" character. (274)

Another state in the mountain west where Bigfoot activity has been reported for the past 30 years is **Colorado**.

In 1970, a trio of elk hunters encountered wildlife of the two-legged kind in the Routt National Forest:

A loud crashing noise behind us came to our attention. My uncle, an experienced hunter and woodsman of many years, replied that it sounded like we had spooked a large bull elk. The three of us started in pursuit.

The animal was obviously spooked and running, as the sounds of its progress through the dense aspen trees and pines were easy to hear. (We) broke out upon the canyon's edge expecting to see the bull elk crossing the canyon bottom below us.

I was alerted to the sound of rolling rocks. As I looked directly across the canyon from me, I spotted what I at first believed to be a very large black bear running up the steep-sided canyon.

Raising my rifle to view the animal through my rifle scope, I was stunned. The creature was running up the steep hill on two feet.

It was covered completely with what appeared to be matted, shaggy, coarse black hair. The full back, shoulders, buttocks and legs were turned to me. The back of the head appeared to be hunched slightly forward, and without apparent neck visible. The back of the head was round.

I did not see the arms or hands, as they were directly in front of the creature and blocked from my view by its body. I believe the creature was using brush and limbs in its path to help pull itself up the incline.

The creature reached the top of the canyon. Without pausing, I saw the creature turn slightly at the hips and peer sideways across at us as it disappeared through the trees. I clearly observed the facial features through the scope.

The face was flat, small-mouthed and thin-lipped. The nose was flat, with large nostril flares. The eyes appeared to be wide apart and set below a low brow ridge. The face seemed to be lighter in color, possibly tan or light brown, and hairless, with a leathery appearance. (275)

In 1987, a man living in Green Mountain Falls (El Paso County) observed an amazing scene outside his home one night:

I looked out the window, and here came these creatures. They were running down the road right in front of my house, which at one point is 30 feet from my window.

They ran with their arms hanging down, swinging in a pendulum motion. They were covered with hair. It was the most incredible thing I've ever seen. (276)

Some of the man's neighbors experienced much closer encounters with Bigfoot around the same time. One man reportedly was knocked down by a tall "thing" while putting out his trash one night A woman said she saw a "large beast" attack her cat. The cat took refuge under her vehicle and survived. Its attacker was last seen running off on two legs. (277)

In 1990, a couple driving near a cabin in Conejos County spotted a black, furry creature standing next to the building. They were sure it wasn't a bear. Later measurement of the awning which the Bigfoot had been standing next to indicated the creature was between 7 and 8 feet tall. The witnesses observed the thing run off on two legs. A human-like footprint 17"-to-18" long was subsequently discovered at the site. (278)

In 1998, four persons observed a Bigfoot pursuing deer in the Roosevelt National Forest (Larimer County):

My son was scanning a mountain with his binoculars when he saw it. My daughter-in-law grabbed the binoculars and looked. She said, "It's chasing some deer. My God, it's got knees!"

She said it ran into a group of pine trees. I got my binoculars and waited. Several minutes later, it leaped out from the trees and went running across the face of the mountain. I swear that thing could run like the wind. It had a very strange, distinctive gait.

I guarantee you it was not a fake. There is absolutely no possibility that it could have been. No human could run that fast across that rugged mountain. It was at least 7 feet tall. (279)

Like the neighboring states of Colorado and Wyoming, reported encounters with Bigfoot in **Utah** date from the 1970s.

In 1977, a group of eight hikers observed a creature covered in white hair engaged in some activity at the edge of a lake in the mountains near North Ogden (Weber County). They estimated the thing's height at 10 feet. It walked off on two legs when a member of the party made a noise by accidentally dislodging some loose rocks. Near where the Bigfoot had been seen, they found the carcass of a skinned and partially-eaten rabbit. *(280)*

That same year, two couples spotted a trio of apparently playful Bigfoot at Elizabeth Lake in nearby Davis County:

We sat on a ridge looking into a meadow 300 to 500 yards away when we saw the first creature. A few seconds later, a second beast—both much larger than humans—entered the meadow. The two romped back and forth in the clearing for at least 10 minutes.

These things walked and ran on only two legs. They had arms, legs and bodies much like a human, but (were) covered with hair. Only their hands and feet did not have hair.

Two of them would romp and play in the meadow while a third stood some 100 yards away at the edge of the meadow near a cluster of pines. When they would run, they would take tremendous strides.

They could not have been bears or other animals. They did not have pointed snouts as bears do, and we had a good, long look at their profiles. At first, we were shocked by the huge size of these things, but then we became intrigued and just stood and watched them.

The witnesses later estimated the creatures they observed were between 8 and 10 feet tall, with shoulders much broader than a human and rounded heads. They added that the Bigfoot remained on two feet at all times, even when crouching. *(281)*

In 1994, a group of Boy Scouts were equally certain the creature they encountered in the woods near Baker Lake (Summit County) was no animal with which they were familiar:

We were walking down a trail at night. It was a full moon. There was a stream, and on the other side I saw a hominid. It was walking on two feet. (It was) not a bear, not a gorilla, not a chimp, not any ape I've ever seen. I'm very, very sure on that.

It was tall, about 6-7 feet. It was all one dark color. It had a high crown. It was built stoutly in the torso, hips and thighs. It was fit, not fat. It moved in a distinct way. It sort of twisted its hips so the outline of the thigh and buttock changed. (282)

Four years earlier (in 1990), a Utah family had to stop their truck to avoid hitting a Bigfoot they encountered one night on a highway near Orangeville (Emery County). The creature stood in the middle of the road for about five minutes some 30 or 40 feet in front of their vehicle as they observed it in the headlight beams. *(283)*

Nevada may seem an unlikely habitat for a creature like Bigfoot, but there were several reported encounters in the Silver State during the 1970s and 1980s.

In 1973, two couples described the strange creature they met on a road near Lake Tahoe:

As we came around a turn, we saw something on the side of the road which we thought was a black bear. As we got closer, it was standing on its hind feet in an upright position.

When it saw our car, it went into the bushes. Just before it disappeared, it turned and looked at us. Its face was flat, like a gorilla's. It was about 7 feet tall, and very shiny.

We went to the sheriff's office and told our experience. They said they had had another report that two girls had seen the same thing and were still crying with fright. The excuse they gave us was it was a crazy person in a gorilla suit trying to scare people.

We don't buy this story, after seeing it. I am a sportsman. I have hunted deer, buffalo, antelope, elk, bear and lion, and I know an animal when I see one. This definitely was an animal. (284)

In 1980, a civilian employee at the federal government's Nevada Test Site (Nye County) reported he came upon an unexpected figure walking along a road in the restricted area. The man was surprised to see anyone on foot at mid-day in such a desolate area. But the figure's appearance was even more surprising:

(It was) *somewhere between 6 and 7 feet tall, standing erect and walking like a man, with dark hair completely covering its body.*

The witness stopped his vehicle and watched as the creature crossed the highway and strode out of sight into the desert. *(285)*

Three years later (in 1983), a Nevada motorcyclist came upon a similar creature one rainy night near Las Vegas (Clark County), as recorded by local police:

He stated that his headlight hit an animal which was standing upright on its hind legs. It turned and looked at him as he approached it, then turned and ran off the road, still upright. He stated that he knew for sure it wasn't a bear.

He stated that the animal ran hunched over, and had long, straight hair all over its body, as well as its face.

The witness estimated the creature was approximately 7 feet tall. A police spokesperson added that the man "was obviously very shook, and seemed quite sane when he came in." *(286)*

Montana has a slightly longer experience with Bigfoot than some of its neighbors.

In 1959, a hunter nearly shot a Bigfoot he encountered near Seeley Lake (Missoula County) before he realized it wasn't a bear. He spotted the creature about 20 yards away with its head and arms resting on a fallen tree.

Just as the man raised his rifle to take aim, the creature looked directly at him and seemed to grin. This was followed immediately by a strange, almost-human scream. As the witness lowered his weapon, the creature stepped from behind the log, revealing a flat head, small ears, short neck and long arms. It then began to sway slowly from side to side, and made a low, rumbling sound. The man promptly retreated from the area. *(287)*

In the mid-1960s, a dozen Boy Scouts found themselves sharing their campsite with an uninvited visitor one night in the Deer Lodge National Forest in Silver Bow County:

The kid next to me heard something going through our backpacks, knocking things over. Then all of a sudden, this thing stepped on my friend. There was a lot of screaming.

We turned our flashlights on it. It was dark brown-black. It walked like a man (and) *took great big strides.* (It) *was three times my size—way over 8 feet tall. We saw it moving away across the ridge.*

There were numerous footprints going across the top of the ridge, bigger than anything I had ever seen before. They were human-like, with five toes (and) *no claw marks. (288)*

A few years later (in 1968), a man spending the night in his camper truck near Billings (Yellowstone County) had an even scarier nighttime encounter with a marauding Bigfoot:

I heard a racket outside the camper. I was going to investigate when the noise stopped. I had left my pipe in the pickup cab, so I picked up my flashlight and opened the door. I was frozen with terror. I was face-to-face with a (Bigfoot).

It had an ape-like face, but it was definitely not a gorilla. The head was slightly pointed, sloping down like the sketches of cavemen. It stood erect, like a man. He must have weighed 600 or 700 pounds and stood 7 or 8 feet tall. The whole body was covered with a reddish-brown hair.

It made a funny noise, sort of like a gargle and whistle at the same time. When those huge, hairy arms reached toward me, I screamed as loud as I could.

The noise must have frightened him, because he jumped back. (He) looked rather puzzled, and he seemed to frown. I leaped back in the trailer, grabbed a .22 pistol and waited for his next move.

I fully expected the creature to come tearing through the door of the camper. He did move forward again, very cautiously, cocked his head in a funny motion and looked through the door. I sobbed with absolute relief when the beast turned away and shuffled back into the darkness. (289)

In 1975, two teenage girls went to see why some horses were acting up at a property near the town of Vaughn (Cascade County). They quickly discovered the cause of the animals' agitation—a figure 7-1/2 feet tall and twice as broad as a man standing about 200 yards away.

One of the girls fired a rifle into the air several times, hoping to scare the creature off. Her effort had mixed results. The creature moved a short distance farther away, but now it was joined by either three or four other similar creatures. The witnesses couldn't be sure of the exact number, because by now they were hightailing it as fast as they could in the opposite direction. *(290)*

The following year (1976), a motorist reported sighting a Bigfoot a few miles away along I-15 near Great Falls. The witness in this second incident saw the creature from about 20 feet away, and described it as between 7 and 8 feet tall and covered with dark brown or black hair. *(291)*

A year later (in 1977), a night fisherman involuntarily shared part of his catch with a Bigfoot along the Missoula River in Lewis and Clark County:

I put some rocks in a circle near the river's edge to put my trout so they would keep fresh. As I approached, I could see some kind of movement, but it was too dark to make out what it was. I switched on my lamp with the beam pointing towards some brush.

Only 10 feet from me, (I) got a full view of a hairy, bipedal creature moving away from me. It was amazing to see how quickly it traversed over the boulders and up the steep embankment. I had the creature in the lamp beam for only a few seconds before it disappeared into the darkness.

I shined my light towards the rock pool where I had been keeping my trout. Before the sighting of the creature, I had four trout in the pool. After the sighting, there were only two. (292)

Around the same time, a young boy riding his horse near Missoula (Missoula County) experienced an up close and personal daylight encounter with Bigfoot:

My horse stopped dead in its tracks as we rode through a meadow. There I was, face to face with a Bigfoot that was about 10 feet away from me. It was peeking out from behind a tree.

It just stood there looking at me and didn't make a sound. I slowly turned my horse around and ran home. It must have stood at least 7 feet or more taller than myself. (293)

In 1982, four teenage boys met up with a Bigfoot twice near the town of Lincoln (Lewis and Clark County):

I felt as if we were being watched. The hair on the back of my neck even stood up. We heard something in the woods ahead of us. The noise stopped, and we continued back towards the cabin. About 400 yards from the cabin, I looked towards the woodline, and I saw a figure standing up and looking at us.

It was black in color, and looked to be at least 6 feet tall. Two days later, we came upon a footprint in the same area. It was in mud and well formed like a human print. I wear a size 12 shoe. I put my foot into the print, and it was a good 4-6 inches longer than mine and and a lot wider.

We then walked back towards the cabin. All of a sudden, my friend stopped and tried to speak. He almost turned pale white and he could not talk. He pointed to an area in the woods.

I finally saw what he saw. Sitting on an old tree stump was what I believe to be a Bigfoot. We were only 50 yards away from him, and I could see clearly. It sat there looking at us.

It looked in another direction while we were staring at it, and it was like he did not have a neck. He moved his whole upper torso, and then looked back at us. When he did this, we took off running. (294)

That same year (1982), some young men got an equine early warning of the Bigfoot they encountered on the Blackfeet Indian reservation (Glacier County):

Our horses started snorting (and) *jumping in different directions like they were scared of something. I don't remember who was the first to sight what we saw. I just heard him swear and ask the rest of us to look.*

We looked in the direction he was pointing, and saw this very tall thing running on two legs beside us. (It) *then turned* (and) *crossed the trail about 40 feet in front of us, and stopped at the edge of the brush. It just stared at us, then disappeared into the trees.*

The best way I can describe it is the hairiest man I ever saw. (It was) *well past 6 feet tall. (295)*

The state in this region with the oldest reported Bigfoot encounters is **Idaho.**

One of the earliest was recorded by none other than President Teddy Roosevelt in his book *Wilderness Hunter*. No precise date is given for the story, but the witness told Roosevelt in 1892 that it had occurred when the then-elderly witness had been a young man. The setting for the grisly events involving the man and his trapper companion was said to have been "the mountains dividing the forks of the Salmon (River) from the head of the Wisdom River." We pick up excerpts from Roosevelt's account at the point where the witness and his partner returned to their camp after a day of trapping:

They were surprised to find that during their absence something, apparently a bear, had rummaged among their things, scattering the contents of their packs and destroying their lean-to. The footprints of the beast were quite plain. His companion remarked:"Bauman, that bear has been walking on two legs."

At midnight, Bauman was awakened by some noise. His nostrils were struck by a strong, wild beast odor, and he caught the loom of a great body in the darkness at the mouth of the lean-to. Grasping his rifle, he fired at the threatening shadow but must have missed, for he heard smashing of the under wood as the thing, whatever it was, rushed off into the forest.

The men kept a roaring fire throughout the (next) *night, one or the other sitting on guard. About midnight, the thing came down through the forest across the brook and stayed there for nearly an hour. They could hear the branches crackle as it moved about, and several times it uttered a harsh, grating, long-drawn moan.*

In the morning, the two decided they would leave the valley that afternoon. All morning they kept together, picking up trap(s)*. They had the disagreeable sensation of being followed. There were still three beaver traps to collect. Bauman volunteered to gather these while his companion went ahead to camp and made ready the packs.*

(Bauman) *shouted as he approached* (the camp), *but got no answer. At first, Bauman could see nobody. Stepping forward, his eye fell on the body of his friend. Rushing towards it, the horrified trapper found that the body was still warm, but that the neck was broken, while there were four great fang marks on the throat. The footprints of the unknown creature, printed deep in the soft soil, told the whole story.*

Bauman, utterly unnerved, abandoned everything but his rifle and struck off at speed, not halting until he reached the meadows where the ponies were grazing. Mounting, he rode through the night. (296)

Although all the signs in the trapper's story pointed to Bigfoot as the most likely culprit, the witness never actually got a good look at the creature involved. Such was not the case when a Bigfoot terrorized the small town of Chesterfield (Bannock County) in 1902, as related in the following contemporary newspaper account:

Residents are greatly excited over the appearance of an 8-foot, hair-covered monster. He was first seen when he appeared among a party of young people who were skating on the river.

The creature showed fight and, flourishing a large club and uttering a series of yells, started to attack the skaters, who managed to reach their wagons and get away in safety.

Measurements of the tracks showed the creature's feet to be 22 inches long and 7 inches broad, with the imprint of four toes. (297)

In 1956, a man reported encountering what he took to be a female Bigfoot when it walked onto a road near Boise. The witness described the creature as 8 feet tall and covered with reddish hair. He said it had a flat face, gave off a strong odor and emitted what he described as a metallic-sounding laughing noise. *(298)*

In 1968, two men panning for gold near the town of Golden (Idaho County) had an experience reminiscent of the one related to Teddy Roosevelt by the old fur trapper—only with a less tragic conclusion.

The pair returned to their camp one day to discover it thoroughly ransacked. Following a trail of flour led them to the now-empty sack taken from their camp. Nearby were impressions in the soil of large, generally human-shaped footprints.

When they returned to camp later that day, they spotted the presumed flour thief—a large, man-like creature covered in cinnamon-colored hair which ran off in "a manner unlike any bear" when it saw them. *(299)*

The following year (1969) saw several Bigfoot sightings in the area around the town of Orofino in neighboring Clearwater County. In one incident, the watchman at a sawmill observed a 6-foot-tall female Bigfoot which he described as having long arms, large breasts, fiery red eyes and an offensive odor. He said the creature was covered all over with dark hair except on its face, hands and nipple area. On another occasion, the same witness reported seeing several Bigfoot creatures accompanied by a large dog. *(300)*

In 1972, a man was riding his snowmobile near St. Charles (Bear Lake County) when he decided to stop and build a fire to warm himself. Suddenly, he got the feeling he was being watched. Looking around, he spotted a creature observing him from about 100 yards away.

The witness described the creature as ape-like and covered with grayish-brown hair. It was standing next to a tree. When the Bigfoot realized the man was looking its way, it reportedly waved its arms, stomped its feet and walked out of view into the woods. *(301)*

Five years later (in 1977), two other men were overcome by a similar sensation that someone was watching them when they stopped for a drink of water at a roadside spring near Steen Mountain. Glancing around the immediate area, the two were shocked to see a large, ape-like creature sitting on a boulder and staring at them. They wasted little time returning to their car and driving away, perhaps in search of something stronger to drink than water. *(302)*

In the early 1980s, an Idaho hunter encountered what may have been a family of Bigfoot near Montpelier (Bear Lake County):

All of the sudden, about 100 yards away, I saw what appeared to be three erect, dark brown beings walking down a trail. They were in order of height, with the tallest leading the way, then a slightly smaller one followed by the smallest of the three, which was trying hard to keep up with the other two.

The leader turned around and waved his/her arm in a manner as if to tell the smallest one to hurry up. They continued to speed walk into the timber and faded away. (303)

Chapter 9

BIGFOOT ON THE PACIFIC COAST

Alaska/California/Oregon/Washington

While the vast wilderness areas of **Alaska** could be expected to offer ideal cover for large creatures, there are for the most part relatively few human residents or visitors to encounter them. Perhaps for that reason, the nation's largest state has produced one of the smallest crops of Bigfoot reports.

The most dramatic of these occurred in the Alaskan panhandle, about midway between Juneau and Wrangell, in the year 1900. It involved a prospector who decided to climb a tree on a hill for a better view in order to gain his bearings. As it turned out, it was one of the best decisions he ever made.

No sooner had the man ascended to a limb perch than he beheld a heart-stopping sight—a group of large, hairy, man-like creatures were running up the hill towards him:

(They were) *the most hideous creatures. I couldn't call them anything but devils, as they were neither men nor monkeys, yet looked like both. They were entirely sexless* (and) *their bodies* (were) *covered with long, coarse hair, except where scabs and running sores replaced it.*

Determined to get away before the creatures reached the tree, the witness quickly slid to the ground and made a headlong dash for his canoe, with the Bigfoot in hot pursuit. He managed to reach it just in the nick of time and paddled off to safety. *(304)*

In 1964, a hunter only saw one Bigfoot about 35 miles south of Fairbanks, but he had reason to believe it may have had company:

Something that I would not believe unless I saw it just stepped up about 30 feet in front of me, stared at me, kind of grunted and walked into the woods real quietly. (It was) *really tall—at least 6 feet—and shaggy, dark brown.*

I headed back to my car, often glancing over my shoulder. I heard cracking branches and low-to-high tones coming from the direction (it) *went. What made me walk a little faster was that those sounds were answered from the other side of the road. (305)*

In the late 1970s, some members of the U.S. military had a longer-range Bigfoot sighting at Black Rapids Glacier south of the town of Delta Junction:

I was part of a group of about a dozen Army personnel training in the area. I was looking across the valley when I spotted movement. It was moving up the valley about a half-mile away.

When it crossed the snow, you could plainly see that it was not a bear. It walked upright with long strides and arms swinging, and moved fast. I have seen bears many times in the same type of terrain, and they do not move like this did. Bears can and do walk upright, usually for short distances when they need to see or smell something and need the height. They don't travel in this manner, and not in difficult terrain.

It was dark in color. It was too big and fast to be a human. The only thing I have seen that looked like this is descriptions of Bigfoot. (306)

In 1992, a couple driving between Anchorage and Fairbanks had a nighttime encounter in the area of Denali Park with a creature that, by process of elimination, may also have been a Bigfoot:

My lights hit something sitting on the yellow line in the middle of the road with its legs pulled up to its chest and its arms folded over its knees. Its head was between its arms, looking toward the ground. It had long, reddish-colored, human-like hair.

At first, I thought it was an orangutan. Then I thought to myself, what would an orangutan be doing in the middle of nowhere in Alaska. I thought the only way that could have been an orangutan is if there's a circus out here. I knew that was not a possibility in such a remote area.

I drove right next to it, and I was at its level. If I had been going slow, I could have touched it easily.

I've lived here almost all my life, and there is no animal native to Alaska that could resemble this thing in any way. (307)

The Canadian province of British Columbia, which occupies the region of the Pacific Coast between Alaska and the lower 48 states, is the scene of the majority of Canada's reports of Bigfoot (known locally as Sasquatch). The Pacific Coast state on the other side of British Columbia—**Washington**—is the state that probably comes to mind most often when people think of Bigfoot. And the reputation is definitely well deserved.

One of the earliest recorded Bigfoot encounters in the Evergreen State occurred near the Cowlitz River in 1917, when a young lumber camp worker walking down a road at night suddenly sensed someone was following him. He decided to slip off to the side of the road and see who it was. The man wasn't prepared for what came walking by.

It was a man-like creature, 6-1/2-to-7 feet tall, walking upright. It had a broad chest and was covered in dark hair.

When he got a look at his stalker, the witness couldn't stifle a loud cry. The sound must've startled the Bigfoot, because it immediately ran off into the woods. *(308)*

One of the most prolonged Bigfoot incidents on record was the one experienced by a party of five prospectors near Mount St. Helens (Skamania County) in 1924. The following verbatim account of the hair-raising adventure by one of the men appeared in a local newspaper at the time:

Six years ago, when we first located our claim, we saw strange, four-toed tracks, with the toes short and stubby and almost square across, and have seen them several times since. Two years ago, the strange animals sought to enter our tent, and we found tracks around our tent in the morning.

About sixteen days ago, Beck and myself saw one of these animals peeking from behind a tree at a distance of about 100 yards. We fired, and I think I hit it in the head. It fell back as though struck. Curious to know the nature of the animal, we crossed an intervening canyon, but it had gone by the time we reached the tree. We noticed large tracks, from 13 to 14 inches long and resembling those of a man.

Wednesday, while we were at the mine, Roy went to the spring for water. As he was returning, he heard a cracking in the alders and turning, saw one of the animals charging out of the woods at him, waving its arms over its head and striking its chest. He fired at it with his pistol, the animal coming within 15 feet of him before it turned back into the woods. He was badly frightened, and hurried back to camp.

Thursday, Roy and Fred encountered another of the animals as they were going to the cabin. They fired at it, and Peterson ran out of the cabin with his revolver. Between them, they fired 16 shots, the final shot by Beck apparently striking the animal and toppling it over the edge of the canyon. This canyon has steep sides, and we could only have gotten down by using ropes.

That night about 11, we were startled out of our sleep by rocks falling through the hole in the roof of our cabin which served as a smokestack. The bombardment continued until after two in the morning. I insisted on building a large fire to frighten away the animals, which pushed against the door and made a great deal of noise around the cabin, drumming on their chests. It was a terrifying experience. We nailed the door shut, and I advised not to shoot so that we would have all our ammunition in case they broke in. In the morning, we were glad to start home.

The newspaper added this editorial comment:

Many have ventured to say that the story is the product of imagination. The miners insist, however, that they saw the creatures. Their stories are given some support from old timers who have heard of the giant creatures that are said to inhabit the country there. (309)

Most people must've found the prospectors' account credible, because the area where the incident occurred is known today as Ape Canyon.

Four years later (in 1928), two men couldn't make heads or tails of the strange creature that visited their campsite in the Olympic Mountains one day:

What we saw was an animal, walking away from us. It was standing erect like a human being. All we could see was its back. We couldn't figure out what it was.

It was very tall and husky, and covered with black, shiny hair that had a few white spots here and there. Its long, hairy arms reached its knees, and hair grew even on top of its hands and feet. It had enormous big feet and human-looking hands, like ours only much larger.

The animal continued to walk away from us. Then suddenly it stopped (and) turned to face us. We got the shock of our lives. It was a male, and it had a human face. It muttered a sound, something like a human voice.

It turned around again, and continued on away from us. As it passed a certain tree, we noticed the apeman was about a foot taller than the first big branch. He stood roughly 8 feet tall.

A little farther on, it stopped and bent over. With one hand, it grabbed a broken tree which lay on the ground—we saw later that the dead tree was approximately 15 feet long and four feet in circumference—put it under its arm, and dragged the darn thing for about 30 feet. Then it dropped it and continued straight ahead, soon disappearing into the woods.

There were a lot of giant footprints. The apeman's footprints were 6-1/2 inches wide by 18 inches long. (310)

In 1957, two hunters couldn't believe their eyes when a Bigfoot made off with a deer they'd just shot at Wanoga Butte. Before they could retrieve their trophy, the 9-foot-tall creature casually walked up to the carcass, picked it up and placed it under a long, hairy arm.

Infuriated by this brazen act of thievery, one of the men shot the Bigfoot in the back several times with his 30.06 rifle. The bullets had no noticeable effect on the creature, which just continued to walk off with its prize, making a "strange whistling scream." *(311)*

Firearms were equally ineffective against an 8-foot-tall Bigfoot with white hair and glowing red eyes encountered by several young men near West Richland (Benton County) in 1966. One of them said he was certain he hit the creature at least three times with shots from a .270 rifle from approximately 20 feet away, while another blasted it from about the same distance with his 12-gauge shotgun. The only reaction from the Bigfoot to all this was a high-pitched squeal. However, the same or a similar Bigfoot took offense when several other area youths threw rocks at it and tried to run it over with a car on another occasion. It scratched the side of the vehicle as it drove past. *(312)*

Around the same time these incidents were taking place, a Bigfoot went on the offensive in neighboring Yakima County. A young man was found lying on the ground, suffering from shock and holding a bunch of white hair in one of his hands. He said he'd been chased and attacked by a white Bigfoot while walking through an orchard. *(313)*

There also occurred in the Yakima area that year the first in what was to be a series of memorable encounters between Washington motorists and Bigfoot. In this one, a driver had to slam on his brakes to keep from hitting the large, white-haired creature with glowing red eyes that suddenly loomed up in his headlights one night. The maneuver caused the man's car to stall just a few feet from the creature. The witness related what happened next:

It just stood there and looked at me through the windshield. He walked around behind my car, turned around and came back, and just bent over and looked in the window at me. I got my car started and took off. (314)

In 1967, there was another case of a Bigfoot exhibiting a taste for fish. One of a group of Native Americans fishing at night on the Nooksack River suddenly felt a strong tug on his net. Aiming a light in that direction, he was astounded to see a Bigfoot trying to steal fish:

It was dark and hairy, over 7 feet tall and its eyes glowed. I never saw anything like it before. It was not a bear.

The man summoned his fellow fishermen in time for them to catch a glimpse of the now-fleeing creature. The next day, 16-1/2-inch-long prints resembling a bare human foot were found in soft mud on the bank of the river. *(315)*

That same year, another Washington man decided to follow a trail of similar footprints he came across in the Cascade Mountains:

I discovered the tracks about a mile from the Little Wenatchee. They were about 15 inches long (and) *seven inches across the ball of the foot. They left an impression indicating considerable weight.*

I began to follow the tracks. When I caught up with the monster, I couldn't believe my eyes.

I brought up my rifle instinctively. I couldn't bring myself to shoot. Not unless I was attacked. That monster looked so much like some sort of primitive human being that I felt I'd be guilty of murder if I killed it except in self-defense.

Instead, the man carefully observed the Bigfoot from a distance of about 75 yards. It was approximately 7-1/2 feet tall, and he estimated it weighed at least 400 pounds. It stood on two slightly bowed legs, and had a pair of long arms which reached nearly to its knees. The creature had dark eyes, a broad, somewhat flattened nose and protruding lips. Its entire body appeared to be covered in reddish fur.

When the Bigfoot spotted the man, it made a loud sound somewhere between a snort and a grunt, then walked off quickly with a gait that reminded the witness somewhat of an elephant's. *(316)*

The following year (1968), a witness who encountered a Bigfoot along a road near the town of Skamania (Skamania County) was confident about his estimate of the creature's size:

Just as I came over this little knoll, there it was, standing in a ditch, over 7 feet tall. I knew the size, because the ditch there is two feet deep and there was a mailbox there, and I know the height of that, and this creature was three (or) three and a half feet taller than the mailbox.

The man was equally certain that the creature he saw wasn't a bear:

You can identify a bear pretty easily. This didn't have a snout. (It was) *brown-to-gray, with a head sort of like a pumpkin* (and) *eyes different than anything I'd ever seen before. (317)*

The next year (1969), it was the turn of a deputy sheriff to encounter a Bigfoot on a Washington road. He was driving near Copalis Beach (Grays Harbor County) in the wee hours of the morning, when he was forced to slam on his brakes to avoid hitting the creature, which was standing in the roadway. The witness got out of his car and used his spotlight to examine the Bigfoot before it walked off into nearby woods:

I would estimate its height at 7-to-8 feet, and its weight at something over 300 pounds. It had hair all over it of a dark brown color. The hair on its head was longer than that on the rest of its body—between five and seven inches long.

It had breasts like those of a woman. They had hair on them also, except for the nipples, which were black like the thing's face. I could see the back of one hand and the palm of another, and I could distinguish fingers. It had legs like a human and buttocks like a human.

When the witness and fellow officers returned to the scene in daylight, they discovered bare footprints that were 18 inches long and seven inches wide. *(318)*

The following year (1970), three teenage girls—one of them the daughter of a Grays County sheriff—said a Bigfoot chased their car between Copalis Crossing and Carlisle, a few miles from the site of the previous encounter.

They described their pursuer as 7 feet tall and covered with silver-tipped hair. A deputy who returned with them to the spot the next day found a series of 16-inch-long footprints indicating whatever made them had a stride of seven feet. *(319)*

In 1972, two other local teenagers had their own nocturnal encounter with a Bigfoot along another Grays County road, this time near the town of Satsop. What made the incident more harrowing for the pair was the fact that they were on foot at the time.

The incident began when the boys heard noises behind them and turned to discover a large, dark figure was following them. They ran to a nearby house, and begged the owner to let them in. The woman was skeptical of their story of a monster chasing them, and went outside to see for herself. What she saw standing about 12 feet from her front porch soon made a believer out of her.

The creature was about 8 feet tall, with broad shoulders, long arms and legs and narrow hips. It was covered over most of its body with 3"-6" grayish-brown hair

The Bigfoot stood staring at the house for a few moments, then turned and walked off into the darkness, easily stepping over a fence in the process. Bare footprints measuring 17 inches long, 8 inches wide at the ball of the foot and 4 inches wide at the heel were found in the woman's yard the next morning. *(320)*

That same year, another Washington law enforcement officer got a chance to observe a Bigfoot up close near Tacoma (Pierce County) while he was sitting in his vehicle on a rural road filling out a report:

Suddenly, I became aware of a strong, offensive odor—very powerful, a combination of rotten eggs, rotten meat and sulphur, but really impossible to explain. Once smelled, though, it is never forgotten. It wasn't a wolverine, a bear, a skunk or any other odorous animal.

I glanced up and saw some shadowy form at the timber's edge. Then to my astonishment, a large, ape-like animal taking huge strides walked directly in front of my parked patrol car—right into the headlight beams.

It stopped, peered intently at me, then strode across the road to a steep embankment on the opposite side. It went up the bank without hesitation taking large, unman-like strides. (321)

Another Washington motorist got an even closer look at a roadside Bigfoot while driving near the Columbia River a few years later:

I kept hearing this slap, slap noise and thought I had a flat tire, so I stopped and checked each tire. When I was checking the right front wheel, with my back to the woods, I could feel something staring at me. I had this feeling of being watched, and I turned around.

That's when I saw Bigfoot. It was the biggest thing I had ever seen alive— 10-to-12 feet tall, with huge, broad shoulders (and) a long torso. It was brown (and) had a huge, hairy body.

He was evidently eating berries or something, because as he chewed I could see his teeth. They were square. His hands were smooth, had no hair inside and were very big. He was like a man, more man-like than animal, and he seemed intelligent. His eyes were enormous, and they were glowing.

I know I screamed, but nothing came out. I don't remember how I got out of there, or even getting back into my car. (322)

In 1975, three campers had the uncomfortable feeling of being followed on a wilderness trail near Rimrock Lake:

We kept hearing these noises, like twigs breaking and brush cracking. We got to camp and heard a noise. I shined the light across the pond, and saw those yellowish-green eyes. Tom shined the light, and we seen the whole body. It was about 8-10 feet tall, big, dark and furry.

The witnesses said it had a human-shaped head, long arms and a thick neck. Just before the men pulled up stakes, they also noticed a disagreeable odor coming from the creature that one of them likened to "rotten loggers' socks." *(323)*

A couple of years later (in 1977), a lone camper also came under scrutiny from a Bigfoot on Mount White Chuck. The woman awoke from a nap and spotted the creature staring at her from about 25 yards away:

(It was a) *black figure, almost human-like, with long, hairy arms and legs and huge shoulders. It was covered with long, black hair which seemed to stand out. He was standing on two legs.*

The woman closed her eyes to pray, and when she opened them again, the Bigfoot was gone. *(324)*

That same year, a Bigfoot proved to be a determined chicken thief on a ranch in Chelan County. It was first seen by two ranch hands who went to investigate noises at a chicken coop. They discovered a 7-foot-tall, gorilla-like creature trying to break in. One of the men shot at the thing, and thought he hit it at least once.

If the Bigfoot was injured by the bullet, it didn't show. The creature ran away, but returned later and made off with several chickens. *(325)*

In 1984, two motorists got a good look at a Bigfoot which crossed the road in front of them one night near Longview (Cowlitz County):

I noticed movement (and) *saw a large, black head and shoulders coming over the guardrail. I came to a stop, turned on the high beams and saw the creature step easily over the guardrail.*

It was about 8 feet tall, with shiny, black hair covering most of its body. As the creature walked, I noticed it never quite straightened its legs, yet it had a graceful stride. Its arms hung to just above the knees.

It was very powerfully built, with (a) *massive chest and arms. Even though it was covered with hair, I could see its muscles flexing in its legs and arms. Most of the hair was an inch or two long, with some maybe four inches on the back and shoulders. This thing was big. In my estimation, it had to have weighed 800 pounds or more.*

As it neared the center of the road, it turned and looked right at us. Doing so required it to turn its upper body, as its head sat low between its shoulders.

I could see its palms as it swung its arms, and noticed it had either no or very little hair in the palm area. Also, the soles of its feet appeared to be devoid of hair. Also noticed was what appeared to be male genitalia. (326)

Around this same time (1984 or 1985), two hunters came upon a vegetarian Bigfoot near Cumberland (King County):

I got a funny feeling that something was watching us, and could smell a strong smell kind of like a wet elk. Then I saw something very big move. It looked like the back of a bull elk, so I gave a loud whisper to my partner. By the time he had walked the half-dozen steps or so from where he was, the thing had stood up. I knew right there and then it wasn't an elk.

As it stood up, it came from a position from squatting between its legs to a full stance. We watched it for a few minutes. It was eating skunk cabbage leaves and roots. It was very wary of us, and kept rocking back and forth. If we made any movements, it would duck its head a few inches and stare at us.

It made a posturing movement like it meant business, so I said let's shoot into the road and see if we can scare it off. So we did, and it did. It took off like a bat out of hell away from us.

The animal was about 8 feet tall or a little more. It probably weighed around 400 pounds. It had grayish-colored hair that was probably 12 inches in length all over its body. It had a blackish-colored face, and almost no neck. Its arms extended below its knees, and were very muscular.

When it took off, it used its arms to propel it to speed, and grabbed at the vine maples in front of it then swung them behind as it took off. It had a small butt, and ran like an ape. (327)

In 1989, a teenage boy stepped in to protect his mother and younger brother when a Bigfoot invaded their property near the small community of Yacolt (Clark County). The incident began when his mother sent him to see why their horse sounded panicky. Running to the corral, he almost ran into the Bigfoot on the path.

The boy ran back toward the house. Looking over his shoulder, he saw the Bigfoot walking away from the corral. At one point, the boy stepped on a branch lying on the ground, making a loud noise. At this, the creature turned and looked at him "with big eyes."

Arming himself with a bayonet, the teenager decided to pursue the retreating Bigfoot, and his mother accompanied him. The pair caught up to the creature, resulting in a five-minute staring match from a distance of about 30 feet.

The witnesses later recorded their observations on the Bigfoot's appearance. It was approximately 6-1/2 feet tall, and covered with 4"-to-5"-long, gray-black hair with golden brown patches on the shoulders and chest. The creature's face had a pronounced brow ridge, deep-set eyes and a short, wide nose. It had large hands with thick fingers at the ends of its long, heavy arms. The witnesses were able to see that the fingernails were yellowish-brown. The mother also noted the creature had "a small butt for a man."

Finally, the mother decided it was time to break off the encounter, and yelled at her son. Whereupon the Bigfoot let out a scream and took a step toward them, prompting mother and son to turn and run home as fast as they could. The Bigfoot did not follow. *(328)*

In 1992, a logger spotted what may have been a female Bigfoot and its young near Mount St. Helens, not far from the scene of the five prospectors' 1924 ordeal in Ape Canyon.

The man first detected an unusually foul odor, then spotted an 8-to-9-1/2-foot-tall Bigfoot. He said it was covered in dark brown or black hair and had noticeable breasts. The witness retreated, but a few minutes later saw the Bigfoot again. This time, it was holding the hand of a smaller creature. *(329)*

The following year (1993), a motorist experienced an uncomfortably close encounter with a Bigfoot while driving near Shelton (Mason County):

It was a clear night and the moon was out, so I could see very good. I noticed something standing in the field. I was slowing down to turn when this thing turned its head and looked straight at me. As I rolled through the corner, it turned in the same direction and walked at the same speed.

It was absolutely huge. It wasn't very far from my car—maybe 40 feet. The field is about 3-4 feet below the road, and this thing towered over me so I had to look over the top of the steering wheel to see it. I could smell it, and it burned my nose.

It took long strides and was actually moving faster than me. I stopped the car, because I thought it was going to come over the barbed wire fence into the road in front of me. It walked down the fence line, and I followed at a distance. It just stepped over the fence with no trouble at all.

I remember feeling scared to death, and the hair on my body was standing up. It crossed the road and disappeared on the other side in the brush. (330)

For a hair-raising encounter with Bigfoot, however, it's hard to top the experience of two off-duty soldiers who were camping near Mount Rainier (Pierce County) one night in 1998.

The pair was awakened in the middle of the night by a strong offensive odor outside their camper that one of them later likened to a combination of a hog barn and rotting flesh. This was soon followed by a fierce growl.

One of the men shone his flashlight through the plexiglass window of the camper—and found himself looking directly into the face of a Bigfoot which bared its teeth meanicingly. Then things really got dicey.

The creature began to rock the truck with such force that the men were sure it was going to roll over. At one point, it seemed the shell of the camper might collapse on top of them. Then, the shaking suddenly stopped as abruptly as it had begun, and no more sounds were heard from the Bigfoot.

Neither of them slept for the rest of the night. At first light, they could see no sign of the creature. The men wasted little time entering the cab of the truck and speeding off. *(331)*

When a Washington forestry worker encountered a Bigfoot near Indianola (Kitsap County) in 2000, he had no trouble distinguishing it from a bear. That's because he was standing between the two at the time.

The incident began when the man suddenly felt the hair on the back of his neck stand up:

I watched this hairy thing on two legs. It used its arm to lift up a branch, and walked about 50 feet. He turned in my direction, saw I was watching him, and ducked behind a tree.

The witness said the creature—which he later described as standing about 9 feet tall and covered with shiny, black fur—began to make a screeching sound and pound on the tree with what sounded like a rock.

Then the man heard another noise from a different spot nearby. When he looked, he saw a bear and its cub about 20 feet away. Interestingly, the mother bear—which would normally be quite concerned about the presence of a human so near her cub—ignored the witness and focused her attention on the Bigfoot. At this point, the man decided that particular part of the forest was getting a little too crowded for his taste and retreated to his truck. *(332)*

While not as closely associated with Bigfoot in the public's mind, Washington's neighbor to the south—**Oregon**—has logged an impressive number of encounters of its own over the years.

Judging from the following 1904 newspaper article, mineral workers were among the first residents of the Beaver State to encounter the creatures:

At repeated intervals during the past 10 years, thrilling stories have come from the rugged Sixes mining district regarding a wild man or a queer and terrible monster which walks erect and which has been seen by scores of miners and prospectors.

The appearance again of the Wild Man of the Sixes has thrown some of the miners into a state of excitement and fear. A report says the wild man has been seen three times since the 10th of last month.

William Ward and a young man by the name of Burlison were sitting by the fire of their cabin one night when they heard something walking around the cabin which resembled a man walking. When it came to the corner of the cabin, it took hold of the corner and gave the building a vigorous shake, and kept up a frightful noise all the time.

Mr. Ward walked to the cabin door, and could see the monster plainly as it walked away. (He) took a shot at it with his rifle, but the bullet went wild of its mark.

The last appearance of the animal was at the Harrison cabin only a few days ago. Mr. Ward was at the Harrison cabin and again figured in the excitement. About five o'clock in the morning, the wild man gave the door of the cabin a vigorous shaking, which aroused Ward and one of the Harrison boys, who took their guns and started in to do the intruder.

Ward fired at the man, and he answered by sending a four-pound rock at Ward's head, but his aim was a little too high. He then disappeared in the brush.

Many of the miners avow the wild man is a reality. They have seen him and know whereof they speak. They say he is something after the fashion of a gorilla and unlike anything else, either in appearance or action. He can outrun or jump anything, (and) *can throw rocks with wonderful force and accuracy.*

He is about 7 feet high, has broad hands and feet, and his body is covered by a prolific growth of hair. (333)

The setting for these incidents was Curry County in the southwestern corner of the state. In 1968, two men camping at Mount Ashland in nearby Jackson County received a less than cordial reception from a Bigfoot in the form of a shower of fist-size rocks raining down on them from above:

We looked up at the ledge and saw this thing. It looked about 10 feet tall, although I realize it only seemed that big because we were looking up at it.

The creature was standing only 50 feet above them, and one of the men used the scope of his rifle to study the Bigfoot in detail:

It was between 7-1/2 and 8 feet tall, and probably weighed about 700 pounds or a little more. It was covered all over with hair its entire body, except for the area around the nose, cheeks and lips. It was a bluish-black hair, pretty long and straight on the head and then growing curly from the shoulders down.

This thing had a pretty big head that went right into its shoulders—no neck at all. I'm sure of that because when it turned to look from one point to another, it moved its entire upper body in order to move its head. It was like the time my brother broke his neck in an auto accident and had it in a cast. He had to move around as if he didn't have a neck when he wanted to shift his line of vision. The thing moved just like that.

It had eyes that appeared close to human. There was a lot of white surrounding what appeared to be a brown iris, with a small black pupil in the center.

It was standing erect. It didn't look like a gorilla or chimp that stands up with its arms dangling and has to support itself on its knuckles when it moves. This thing was standing like a human being. It moved like a human being.

Its shoulders were enormous. I'd guess they were close to four feet across. The body tapered down only slightly into a waist. Then it flared out again in large hips and very large buttocks.

As I watched it through the scope, I was thinking to myself that I was glad I hadn't taken a shot at it. Because that thing was human. Or close enough to it to make me happy I didn't shoot it.

It stood there for a couple of minutes, just watching us. Then it let out a weird sound. It wasn't like anything we had heard before. The best I can describe it is that it was a high, shrill whine. Something like the screech you get when you overload a tape recorder or accidentally put the line jack into the mike connection. But it changed tone and it changed pitch.

Then the Bigfoot turned and climbed the rest of the way up the hill effortlessly. When the two men followed the route it had taken, they found the going much more difficult. Their efforts were rewarded when they discovered fresh bare footprints on top of the hill:

They were human in form—five toes, with the big one sticking out like on a human foot, not turned for gripping like a monkey's foot. This print was 18 inches long and about six inches wide at the ball.

The fantastic thing was the length of its stride. It was maybe 60 inches long—twice as long as mine when I walk fast. That was the average stride of the couple of dozen prints we saw. They ranged between four feet and six feet long, as if the thing was hurrying, then slowed down, then started to move quickly again. (334)

Two years later (in 1970), a young boy got a briefer but all too close look at a Bigfoot one evening near the town of Sheridan (Yamhill County):

The first thing I noticed was the smell. It was terrible. (If) you ever had your dog roll in something that was dead, that's exactly the smell—the type that sticks in your nostrils.

That's when I turned on my flashlight. There, in front of me, standing on two legs, was a (Bigfoot). It was huge—around 7 feet. The head looked somewhat pointy, with shaggy hair hanging from it. The cheeks had less hair. The jaw protruded something like an ape's, but was more blocky on the chin than an ape.

I couldn't see a distinguishable neck. The head seemed to set directly on its shoulders. Its shoulders were quite broad, maybe four feet or better. The torso and hips were somewhat less broad.

The arms seemed longer than a human, a few inches above the knees. The hair on its body (ranged) from three or four inches to very long. The eyes really caught my attention. In the flashlight beam, they glowed back like red embers. (335)

About the same year, a teenage girl had a face-to-face confrontation with a female Bigfoot who was using a five-fingered discount to shop for meat at a logging camp near Detroit Lake (Marion County). The witness had been awakened by a noise on the porch in the middle of the night and gotten up to investigate.

Clearly illuminated by the light from an open refrigerator door stood a 6-1/2-foot-tall creature with reddish-brown fur and breasts. It had a 20-pound piece of meat tucked under one arm.

The two stared at each other for a few moments, and then the girl screamed. By the time several men from the camp arrived on the scene, the Bigfoot and the meat were gone. The creature left behind 14-inch footprints. *(336)*

In 1971, a man had a chance to shoot a Bigfoot he and his wife encountered one evening near The Dalles (Wasco County). Their car headlights had revealed the creature standing in a field next to the mobile home park where they lived, and the man went inside for his rifle. He then observed it through the eight-power scope for about five minutes from a range of 150 yards.

The man estimated the Bigfoot was close to 10 feet tall, and weighed somewhere between 600 and 800 pounds. It was covered in hair and looked very muscular. He couldn't bring himself to shoot the creature, he said, because "It seemed more human than animal."

The Bigfoot eventually walked out of view, and the witness called the sheriff's department. Deputies later found 20-inch-long prints resembling a bare human foot where the creature had been seen. *(337)*

The following year (1972), several other Oregon men were less hesitant about pulling the trigger on a Bigfoot near Millersburg (Benton County). They responded when a woman reported a 6-1/2-foot-tall "man" completely covered with hair prowling her farm. Two men pursued the intruder and fired several rounds at it with shotguns. Police later found a fresh pool of blood at the scene. *(338)*

A couple of years later (in 1974), two teenage boys were packing considerably less firepower when they encountered a Bigfoot near the town of Sandy in nearby Clackamas County:

We were shooting our BB guns at a stump a little ways down hill from where we were. Something with a human-like head passed right in front of the stump. The trees and brush obscured everything but the head, but I got a good look at it. It had brown fur over almost all of the head a little darker than cinnamon color (and) *a flat nose.*

It walked by and disappeared into the brush. We went down to where it passed, and figured it had to have been about 7-1/2 feet tall. (339)

In 1980, a motorist got a full-length look—and a noseful—of a Bigfoot on a road between the towns of Bend and Sunriver (Deschutes County) one evening:

I saw what I at first thought was a deer come from the brush and trees on the side of the road. I slowed down for it to cross in front of me. It was not a deer, though. It was a very large, light reddish-brown, very hairy man.

When the car lights hit his eyes, he stopped in front of me and shielded his eyes. I had what we called running lights on the car, so he was lit up like a Christmas tree. I stopped the car, because I was afraid I would hit him. He was looking face to face with me.

His skin was a light brown or tan, and he had freckling on his hands and exposed areas around his eyes. He (had) *a neck, but it was very short. The muscles in the top of his arms and shoulders were built up like a weightlifter's. His teeth were brown, his nose* (was) *full and wide, and his mouth had full lips. I know he was male, because I could see his male parts*

He had a very strong smell—like a man who had been hitchhiking for a few weeks and a dog mixed together. The smell stayed in the car for a while.

He turned and walked off the road. (340)

Later that decade (in either 1986 or 1987), a teenager camping on Mount Hood (Hood River County) believed he may have witnessed a Bigfoot giving parental guidance to its young one morning:

Some movement about 70 feet away in the berry bushes and evergreens caught my eye. I saw a large, light beige-colored creature, all covered with hair, 7-to-8 feet tall, its back to me.

Not more than 10 feet away was this other creature, the same but small, all covered with hair except for the front of the hands, the bottoms of the feet and around the eyes. The second one was only about 3 feet high (and) *a bit darker in color. The hair on both was up to four inches long.*

The big one was really thickset. I could not make out any of the front of "her," because she was turned away from me. I thought she was the little one's mother. (The little one) *was picking up a stick which it was trying to put in its mouth. She gave a kind of grunt at the little one like she didn't want him doing that, and he dropped the stick.*

At that moment, I was on all fours leaning out of the tent, trying to see better, and my hand popped on a twig. The big one looked right at me. She grunt(ed) *again at the little one, and reached down and took his hand. She looked in my direction one more time, grunted softly, and they were gone behind the trees. (341)*

What appeared to be a pair of playful Bigfoot were observed near Wallowa (Wallowa County) in 1988:

I was walking down a dirt road at daylight when I seen some movement. At first, it almost looked like two bears standing up on their hind legs facing each other.

I put my binoculars on them and witnessed what seemed to be two young Bigfoot playing. I watched them jump around chasing each other, and jump up and hit their hands together for about a minute. Then, all of a sudden, they both just stopped and walked off together into the trees.

They were both about 5-1/2 feet tall, with hands down almost to the knees (and) *long brown hair. (342)*

In the late 1980s, two fishermen were walking near Linton Lake in Lane County when they surprised an animal they at first thought was a bear. But when they got closer and the creature stood up and faced them, they saw it was a 7-foot-tall Bigfoot.

One of the men started to reach for a .22 pistol, but his companion yelled at him to stop. At that, the Bigfoot leaned down, picked up a deer carcass from the ground, slung it over its shoulder and walked off into the woods. *(343)*

In 1993, a convoy of soldiers on a military training exercise observed a trio of Bigfoot creatures near Saddle Mountain (Clatsop County). One of the witnesses described them as "large, bipedal hominids covered with black or very dark brown hair from head to foot." He said one creature was 7 or 8 feet tall, and the other two were about a foot shorter. All three of the creatures stood erect for the duration of the sighting. But while the big one remained perfectly still, the others continually shifted their weight from one foot to the other, producing a rocking motion. Another soldier who observed the creatures walking said they had an exaggerated arm swing, and moved with a graceful stride that "ate up quite a bit of real estate." *(344)*

Two years later (in 1995), an Oregon bow hunter became the hunted when he unwittingly attracted the attention of a Bigfoot near Colton (Clackamas County):

I heard crunching footsteps coming from directly behind me. At the time, I thought it might be a buck in rut. A scent had been applied to my clothes. The animal seemed to be following (the) *scent directly to where I was hidden in some blackberry bushes.*

To the man's chagrin, the animal he'd managed to entice with the scent of a female deer wasn't a male deer, but rather a Bigfoot:

I estimated it to be about 7-1/2 feet tall, and maybe 600 pounds. It seemed to know exactly where I was sitting. We were staring at each other from a distance of about 75 feet for a full minute. The Bigfoot swayed back and forth a few inches from side to side. Eventually, (it) *turned around and walked in the same direction it came from. (345)*

In 1996, a couple driving near LaPine (Deschutes County) encountered a Bigfoot who was obviously convinced it had the right-of-way:

I saw a tall, rangy figure suddenly emerge from the cover of the treeline about 150 feet ahead and stride determinedly toward the road. As I slowed and closed the distance to within 50 feet, the figure—with no hesitation (and not) *looking in the direction of my car—glided across the two lanes in four strides and kept going directly into the cover of the pines on the other side of the road.*

The figure seemed very tall. I would estimate close to 7 feet, based on (its) *height relative to the roof of my car. The length of the stride required to cross the road in just four strides would be quite large. It moved briskly, with a long, swinging arc of the arms* (and) *a slightly stiff body lean, with the head and trunk bent forward from the waist. The head was not bent at the neck. I saw no neck. The whole upper torso was bent from the waist. (346)*

In 1999, two Oregon college students were investigating a supposedly-haunted abandoned house near Medford (Jackson County) when they scared up a living nightmare instead:

We heard a rustling in the woods next to us about 50 yards away. The rustling began to get closer, but we thought it was a deer or something.

All of a sudden, there was this loud, deafening scream-roar-growl noise, and we heard what sounded like heavy footsteps in the field next to us. I saw, running at us, what appeared to be a massive creature with long, shaggy hair. It had to have been at least 8 feet tall.

It stopped about 25 feet away. We took off running for the car, and the creature pursued us. By the time we got to our automobile, another car was coming down the dirt road, and it scared the creature away. (347)

Bigfoot's apparent ability to adapt to a wide range of climate and terrain conditions is perhaps nowhere better in evidence than in **California**, where frequent reports of the creature have come from the state's mountains, flatlands and desert areas.

Among the earliest recorded Bigfoot encounters in the Golden State is the following one by a man hunting in the northern mountains in 1869 and reported in an area newspaper shortly after the incident occurred:

Several times I returned to my camp and saw that ashes and charred sticks from the fire had been scattered about. I was anxious to learn who or what it was that so regularly visited my camp. In damp sand, I (saw) the track of a man's feet, bare and of immense size.

Now I was resolved to lay for the bare-footed visitor. I took a position on a hillside about 60 or 70 feet from the fire, and hid in the brush. I waited and watched.

Suddenly, I was startled by a shrill whistle. I saw the object of my solicitude standing beside my fire, looking around. It was in the image of a man, but it could not have been human. I was never so benumbed with astonishment before.

The creature, whatever it was, stood full five feet high, and (was) disproportionately broad and square at the shoulders, with arms of great length. The legs were short and the body long. The head was small compared with the rest of the creature, and appeared to be set upon its shoulders without a neck. The whole was covered with dark brown and cinnamon-colored hair.

As I looked, he threw his head back and whistled again, then stooped and grasped a stick from the fire. This he swung around and around until the fire on the end had gone out. Then he repeated the maneuver.

Fifteen minutes I sat and watched him as he whistled and scattered my fire about. I could have easily put a bullet through his head, but why should I kill him? Having amused himself apparently all he desired with my fire, he started to go. Having gone a short distance, he returned and was joined by another, a female unmistakably.

Then the both turned and walked past me within 20 yards of where I sat, and disappeared into the brush. There only object in visiting my camp seemed to be to amuse themselves with swinging lighted sticks around. (348)

In 1897, there occurred the first of several related incidents which, if authentic, would surely rank as the most remarkable case of human-Bigfoot bonding on record. They are alleged to have occurred at Tulelake (Siskiyou County), near the Oregon state line. The story was related by the participant, a Native American who steadfastly maintained to his dying day that the events were true. The following account was recorded by the man's grandson:

He was walking along a deer trail near a lake just about dusk, when he saw up ahead something that looked like a tall bush. Upon coming a little closer, he became aware of a strong odor, sort of musky. He then gave a close look at the bush, and suddenly realized that it was not a bush at all, for it was covered from head to foot with thick, coarse hair.

He took a step closer, but the creature made a sound that sounded like "nyah." Grandfather was able to see quite clearly two soft brown eyes through the hairy head part. Then the creature moved slightly, and grandfather laid down the string of fish he had been carrying. The creature quickly snatched up the fish and struck out through the timber. It stopped for a moment and made a sound, a long, low "agoum."

A few weeks after his encounter, he was awakened one morning by some strange noises outside his cabin. Upon investigating, he found a stack of deer skins, fresh and ready for tanning. Off in the distance, he heard that strange sound once again—"agoum." After this, there were other items left from time to time, such as wood for fuel, wild berries and fruits.

It was a few years later that grandfather had his second, but far more amazing contact with the Sasquatch. He was struck on the leg by a timber rattler. Grandfather soon found it difficult to go on, became sick at his stomach and fainted. When he came around again, he was surrounded by three large Sasquatch about 8-to-10 feet tall. They had made a small cut on the snakebite, and somehow removed some of the venom and placed cool moss on the bite.

Then one of (them) *made a kind of grunting sound, and the other two lifted him up and took him down the mountainside, placed him under a tree and left. (349)*

In marked contrast to these incidents, Bigfoot was suspected of totally opposite behavior towards humans at the opposite end of the Golden State. Deadman's Hole near Warner Springs in San Diego County got its grisly name from the fact that several battered remains were found there in the 19th Century with their necks broken, apparently strangled by powerful hands, and their bodies drug around.

When two more people became victims in 1922, a pair of hunters resolved to track down the killer, which they suspected was a transient bear. The creature they trailed and shot, however, was no bear. According to one of the men:

That thing looks like a gorilla. I got a close look. The hands and feet were really big. The face was almost human.

The witnesses added that the creature's head was rather small for its body. They gave no estimate of its height, but guessed it must have weighed about 400 pounds. Its feet were reportedly a whopping 24 inches long. There's no record of what became of the creature's corpse, which was presumably left or buried near where it was killed. *(350)*

In 1962, a man ran into a Bigfoot near the town of Fort Bragg (Mendocino County)—literally. He rushed outside a friend's house to see why the family dog was barking so fiercely and collided with a large "half-man, half-beast."

The terrified man jumped to his feet and sprinted into the house with the creature hot on his heels. The Bigfoot began pushing on the door before the witness could close it all the way. The man and his friend then engaged in a contest of strength with the creature from opposite sides of the door. It was a standoff, with the men unable to close the door all the way and the Bigfoot unable to open it farther.

Then the Bigfoot slacked off its pressure, and the home's owner ran for his rifle. By the time he returned, the creature was gone. *(351)*

The following year (1963), a camper preferred not to think how close a Bigfoot may have been to him in the Sierra National Forest:

I decided to take a short nap. I awoke with the odd sensation that I was being observed.

There, in front of me and not over 10 feet away, was an "individual" of considerable bulk and stature, rapidly striding away from me.

What I had seen was a creature covered with grayish-black hair, whose head had a pointed quality toward the rear, not rounded as a man would have. Any skin about the face was the same color as the fur covering the head.

He had the appearance of muscular bulk, like one would expect to see at Muscle Beach. His height had to have been at least 6'6".

His stealth in the woods was remarkable. The ground was covered with branches and debris from fallen trees, but he made no sound, and moved very rapidly away from me. (352)

A couple of years later (in 1965), a Bigfoot put a premature end to a cookout for a man and wife near Hesperia (San Bernardino County). The couple was roasting hotdogs over a campfire after dark when they heard a noise nearby. Looking around, they spotted a pair of glowing (or reflecting) eyes observing them from the darkness. More ominously, the eyes belonged to a hairy, man-like creature which stood 9 or 10 feet tall, with long arms and no neck.

The two quickly gathered up their things and headed for their car, which was parked nearby. It was fortunate for them that their vehicle wasn't parked farther away, because the thing was following (walking upright like a man) and gaining on them. It had almost caught up to them when they hopped in their car and drove off. *(353)*

The Bigfoot may have acquired a taste for the hotdogs the witnesses left behind, because a few months after this incident, a nearly identical scene was played out in the same area when an 8-foot-tall creature with reflective yellow-green eyes and a head "the shape of a football helmet" suddenly approached to within a few feet of another couple's campfire. This second couple wasted little time returning to their vehicle and departing the area. *(354)*

Another Bigfoot may have been looking for food when it was discovered rummaging through the parked car of five people who were picnicking in a forest near San Diego in 1966. The creature growled and ran off when the witnesses approached it, but then lunged at their car from behind a tree as they drove away.

A little farther down the road, they saw a creature similar to the first seated beside the road. It just looked at them as they drove by. The people said both creatures appeared to be well over 6 feet tall and covered with reddish-colored hair. The witnesses were unanimous in their conviction that the creatures were not bears or any other animal with which they were familiar. *(355)*

Arguably the most famous Bigfoot encounter of all time occurred near Bluff Creek in Del Norte County in 1967. It involved two men traveling on horseback, one of whom carried a movie camera loaded with 16mm color film. Here's how the photographer later recounted the incident:

We were packing our horses back into one of the last remaining great wilderness areas northeast of Eureka. We rounded a sharp bend in the sandy arroyo of the creek. Then it happened.

The horses reared suddenly in alarm and threw both of us. Luckily, I grabbed by camera. About 100 feet ahead, on the other side of the creek bed, there was a huge, hairy creature that walked like a man. It was a female.

We estimated later, by measuring some logs that appear in the film, (she was) about 7 feet tall. I estimate that she would weigh about 350 pounds. She was covered with short, shiny, black hair—even her big, droopy breasts.

She seemed to have a sort of peak on the back of her head, and she had no neck. The bottom of her head just seemed to broaden out onto and into her wide, muscular shoulders.

She was just swinging along, but all of a sudden, she just stopped dead and looked around at me. (356)

The film the man took that day runs for only a few seconds. But the controversy it generated is still running. Some people look at the footage and see muscle movement and other physical features they say could only denote a living, non-human animal. Others view the same film and are just as certain it's a man in a fur suit. (A few even insist that photo enlargement reveals a zipper down the back.) Individuals have claimed to know who constructed the suit or the identity of the person who supposedly wore it that day. To date, there's been no conclusive resolution to the controversy one way or the other.

In 1968, a prospector said he encountered a "sandman" in the desert near Borrego Springs (San Diego County), not far from the scene of the deadly doings at Deadman's Hole years earlier:

I was camped on a mesa one morning when I saw a man walking in the desert. The figure came closer. I thought it was another prospector. Then I picked up my binoculars, and saw the strangest thing in my life. It was a real giant apeman.

That thing was big. I was no match for it. I had a .22 pistol, but it would have been like shooting a gorilla with a peashooter. I was afraid the beast might get too close. So I fired a couple of rounds into the air.

The sandman jumped a good three feet off the ground when the sounds of the shots reached him. He turned his head, looked toward me and then took off running in the other direction. (357)

In 1970, three teenage campers claimed considerable interaction with several Bigfoot in the Trinity National Forest (Shasta County). Their first contact was with a creature they spotted observing them from a hill one evening around dusk. When one of the boys started walking toward the Bigfoot, it threw a rock at him which landed about six feet away. Then it turned around and walked off.

Later that night, the female member of the trio was awakened by noises outside the car she was sleeping in. She peeked out and saw a large, dark figure standing next to the vehicle. The girl remained perfectly still, almost afraid to breathe, hoping the creature would go away. Eventually it did.

Around dusk the following evening, they spotted a Bigfoot about 75 feet from their campsite. When one of the boys approached it, the creature extended an arm outward from its hip. The boy responded with a similar gesture, and the two of them then exchanged the same movement three or four more times.

Next, the Bigfoot assumed a low crouching position, and the boy responded with a reciprocal movement. Apparently tiring of the game, the creature then stood up, turned and walked away.

The witnesses described the Bigfoot as between 8 and 10 feet tall, and thought it must have weighed somewhere between 500 and 800 pounds. It had an egg-shaped head with a hairless, dark-skinned face and small, round eyes set deeply below a heavy brow ridge. Its long arms hung about halfway down its thighs. *(358)*

A couple of years later (in 1972), Shasta County was also the setting for a cat-and-mouse game between two campers and Bigfoot near Clear Creek. The contest began after the young men had taken refuge under a bridge during an early morning downpour:

(We heard) *thump, thump, thump coming across the bridge. Whatever it was sounded heavy, and tossed rocks over the side.*

Suddenly, there was a noise across the river. Randy shined the light over there. We could see this thing standing there.

One of the men fired a pistol in the direction of the creature, and it ran off. But not for long:

With daylight, we could see this creature on the hill, looking down at us. It moved from tree to tree, watching us. It was definitely not an ape, because it was too much like a man. When it ran, it bent its knees.

(It was) *almost like he wanted us to track him. We tried to follow, but he was really smart. He'd stay on the rocks so we couldn't see footprints. Just when we'd think he was gone, just then he'd be again, looking at us from behind a tree.*

The men eventually conceded the game to the Bigfoot and returned to their camp. They later described their adversary as being completely covered in thick, rusty brown-colored hair. They weren't able to form a good opinion as to its height. *(359)*

That same year (1972), a family with a home in the mountains east of San Diego found out at least some Bigfoot have a fondness for fruit when they were plagued by "big monkeys" picking apples from their trees. Family members observed as many as three of the creatures in their orchard at a time, usually at night.

The witnesses described the thieves as hairy, heavily-built creatures which walked on two legs and resembled a cross between a man and an ape. The nocturnal forays ceased when the family installed large floodlights. *(360)*

It was also around this time that a babysitter near Palmdale (Los Angeles County) got a real-life scare worse than in any horror film. She first noticed the family dog barking and whining outside the house one night. When the young woman went to investigate, she was confronted by something that didn't need a mask to be frightening:

(It was) *a creature 8-to-10 feet tall, hairy all over, with a wild look in its eyes—a creature that was more man than animal.*

In the brief interval the creature stood towering over the terrified girl before she ran back in the house, she was nearly overcome by the "awful" odor emitted by the beast. *(361)*

Two other teenage girls had a scary Bigfoot moment of their own in the Kern Valley (Kern County) in 1973 or 1974. They were sleeping in a small trailer, and two brothers of one of the girls were sleeping in another trailer nearby.

In the middle of the night, the young women were suddenly shaken awake by the violent rocking of their trailer. Suspecting a prank by the two boys, the witnesses opened the curtains with the intention of registering their displeasure at the timing of such shenanigans.

To their dismay, they found themselves staring into a hideous-looking face with glowing red eyes set on broad, hairy shoulders. Based on the height of the window, the creature must have stood about 8 feet tall. Mercifully, the Bigfoot then disappeared as suddenly as it had arrived. *(362)*

In 1975, it was a group of Boy Scouts who were on the receiving end of a Bigfoot shock treatment in the San Bernardino National Forest (San Bernardino County):

We were all gathered around the outdoor stove telling jokes and stories when we heard the first scream. It started out like a wolf howl, but ended like a man's scream. It stopped the crickets from chirping. Our scoutmaster came out of his tent to listen as well. Then we heard it again. Needless to say, we were pretty scared.

We all went to sleep. My buddy Kenny woke me up and told me that there was something moving around in the bushes and trees next to our campground. We all woke up, and Kenny walked over to where the rustling was coming from.

He shined his flashlight, and there it was. A (Bigfoot) was standing in the treeline, looking at us. We all broke out yelling and screaming.

It was about 6-7 feet tall, and had brown hair all over except the face. It had a very short neck, and long arms going down to the knees. It's eyes reflected the flashlight's beam. The eyes shone red.

It turned, and with long steps, walked up a steep wooded hill and out of sight. (363)

Two years later (in 1975), Bigfoot was seen again in one of its favorite stomping grounds, the area around Mount Shasta (Siskiyou County). The witness this time was a veteran forestry worker who heard someone moving toward him through the woods. Then a creature unlike any he'd ever seen before appeared from behind a bush and stared directly at him from a distance of about 25 feet:

I knew it wasn't a bear. Bears don't walk through the woods on two feet. It had to be about 7 feet tall, but I don't know what it was. It smelled rancid and rotten, like an old bear hide. That's when I took off. (364)

It was also during the 1970s that Bigfoot staged a number of unauthorized nighttime entries at Edwards Air Force Base in Kern County. A member of the installation's security force who observed the strange intruders through a powerful night vision telescope on more than one occasion said he saw as many as five of the creatures at once. Most were between 6 and 8 feet tall, but he once saw an individual that was over 10 feet tall. Another time, the airman spotted what he took to be a female Bigfoot walking with its young. The witness was able to study the creatures closely, and said they were covered in hair everywhere except their faces, palms and soles of their feet. Their faces resembled apes, with small eyes, flat noses and ape-like lips. They had long arms which reached to their knees.

In one incident when a motion detector indicated an intruder, a base officer and an airman both observed a tall, hair-covered apeman walking across the desert terrain. The creature took off running when it was spooked by the noise of a low-flying helicopter. *(365)*

In 1985, two men were harassed by Bigfoot while hiking in the desert just east of the same base. First, rocks were thrown at them by an unseen assailant, and then they were followed by something making peculiar screams. Finally, they glanced behind them and spotted two creatures about 75 feet away:

One was 9 feet tall and black, the other 8 feet high with dark fur and a tan face. No one who has seen them and a bear would mistake one for the other. (366)

Two years later (in 1987), three loggers had their evening meal interrupted by Bigfoot at a picnic area in the Tahoe National Forest. The creature announced its arrival with a series of "screeching, squawking noises," then emerged from the trees walking in their direction. One witness described it as between 9 and 10 feet tall and covered with black hair.

When the Bigfoot saw the men, it altered its course and ran off, taking five foot strides. But by this time, the witnesses had already lost their appetites:

Two strides was enough for me. I packed up my grub and got the hell out. (367)

A 1988 experience of two hunters in Madera County near Yosemite National Park illustrates just how human-like Bigfoot can sometimes appear:

Early in the morning before light, my friend and I decided to hike up a ridge where we could see a lot of area, hoping to see a deer that we could bag. We parked in an area just below the ridge.

As it got light enough to see the top, we noticed someone already on the ridge wearing what seemed to be black of all things. He had no light and no gun, and we wondered where he came from, because we had been camped there for 11 days and no one had come by that whole time.

We watched him dig around in some brush for at least 30 minutes, pausing occasionally for long periods to just stare in different directions. We joked that it was probably a hunter from L.A., because anyone would be crazy to wear just black during deer and bear season.

Then he began to walk down the hill right for us. He would walk a little ways and then stop and look around, just as if he was hunting. He walked just like a human, so at this time there was no reason to think it was anything but a man.

When he was about 50 yards away, we both realized we were not watching a man. This was a Bigfoot! He had jet black, shiny fur, with a reddish color around his eyes.

I tried to open the door of the truck real easy so he wouldn't notice, but he heard the latch release. (He) *took one good look before breaking into a run. (368)*

In 1992, two young boys observed a "big, hairy man" close enough to smell him in the forest near Klamath (Del Norte County):

The smell was like rotted chicken. It was awful. He was covered with thick, dark brown hair, and was shaking a branch in his hand. We could see his face real good. (369)

The following year (1993), a Bigfoot once again demonstrated a fondness for rattling the cage of people inside parked vehicles, this time near Big Bear Lake in San Bernardino County:

I was sleeping in my van and my two friends were sleeping in a tent about 15 feet away. I heard something walking around outside.

All of a sudden, the van started rocking violently back and forth, and I was thrown from side to side. I yelled real loud, and the shaking stopped. I grabbed my flashlight, pulled back the curtains covering the back window, and shined the flashlight outside. A figure moved. It was walking upright like a man, but it had hair from head to toe. I saw it walk into the forest.

I jumped out, yelling for my friends to get up. It smelled really bad, like a cesspool, around our campsite. We talked about what had happened, and I learned that something had been tossing pebbles onto the top of the tent and had brushed up against the back of the tent. They thought it was me. We decided to pack up and head home early. (370)

In 1998, a Bigfoot also provided some anxious moments for six people camping in the mountains near Hayfork (Trinity County) after it put a damper on their marshmallow roast one night with its bloodcurdling screams.

When one of the group shone his flashlight in the direction of the noises, he revealed the culprit. A 9-foot-tall Bigfoot with yellow eyes and arms hanging past its knees was standing on the opposite side of a creek.

As far as they know, the creature never approached any closer to the campsite, but it kept them awake the rest of the night with its screams. In the morning, the campers found footprints 20 inches long and six inches wide. *(371)*

Around the same time, four people watched in disbelief as a Bigfoot appeared to go for an evening swim (or else commit suicide by drowning) at Little River State Beach near McKinleyville (Humboldt County):

(My wife) *brought my attention to a man strolling at a pretty good clip from the direction of the highway towards the ocean about 30 feet away from us.*

This guy was huge (and) *covered with hair. We all agreed he must have been about 7 feet tall or better* (and) *must have weighed better than 600 pounds.*

He was hell bent on getting somewhere fast, and the only place in front of him was the blue Pacific. Sure enough, we watched him charge out into the ocean and disappear into the darkening waters.

We took a flashlight and went to take a look. The tracks in the sand must have been two of my feet long and some wider. (372)

Farther south along the California coast, a woman and her two children visiting from the Midwest were looking for wildlife at Montana de Oro State Park in San Luis Obispo County in 1999 when they spotted a totally unexpected specimen of local fauna looking back at them:

I raised the binoculars and could see at the top of the hill a huge, black figure standing there, seemingly looking right at us.

The edge of the body appeared jagged, like fur. It had massive shoulders, and seemingly no neck. Then, it turned and walked back down the other side of the hill.

We could also see clearly through the binoculars hikers going up the side of a hill (near) *where we saw this creature. We could clearly see their skin, and could tell they were much smaller than the creature, which seemed to be 8 or 9 feet tall. (373)*

Chapter 10

OTHER EVIDENCE

While the accumulated testimony of eyewitnesses is impressive in itself, it's not the only proof of Bigfoot.

The type of evidence most often left behind by Bigfoot is footprints. Thousands of prints that are generally human-shaped but much larger (sometimes measuring well over 20 inches long) have been found from coast to coast, frequently in conjunction with a creature sighting.

Included are prints appearing to show anywhere from three to six toes, usually with no indication of claws. Some of these footprints have undoubtedly been the work of hoaxers. In 1982, for example, a retired forestry worker claimed he carved a pair of wooden, foot-shaped molds and faked Bigfoot prints in the snow in Washington state in 1924. *(374)*

However, several scientists who studied casts made of more recent prints concluded most of them showed unmistakable signs of having been made by a living animal. Their microscopic examinations revealed such features as dermal ridges (similar to whorls seen in fingerprints) and sweat pores, as well as occasional deformities and indications of injury. One police forensics specialist said the dermal ridges he examined resembled neither human or ape, and would have been impossible to fake. *(375)* According to another expert, the prints he studied indicated the creatures which made them had ankle bones located farther forward than humans, an anatomical detail he said was consistent with supporting a body weight of up to 800 pounds. *(376)*

Samples of what is believed to be Bigfoot hair have also been the subject of scientific examination. The results generally showed them to be either from a non-human primate (but not a known one such as a gorilla or monkey) or somewhere between human and animal. As one expert concluded:

It's really quite unique. It doesn't fit any specific pattern of human or animal hair. It sort of overlaps. (377)

Analysis of another hair sample was judged to be "a mixture of human and animal characteristics." *(378)*

Several of the Bigfoot encounters described earlier in this book featured anomalous hair samples left at the scene. For example, analysis of hair found after one of the creatures made a shambles of that home in Carter County, Oklahoma in 1983 could not link it to humans or any known animal. *(379)*

Hair samples retrieved by the father and son turkey hunters who were terrorized by two Bigfoot creatures in Hampshire County, West Virginia in 1986 were subsequently analyzed by a research center in the Washington, D.C. area, which released a cautious statement on the results:

We can only conclude that they came from some primate species. The hair did not come from a gorilla or from one of the more common species of monkey. It did not come from a human or non-primate. (380)

And remember that Maryland motorist who collided with a Bigfoot in 1975? Laboratory analysis of hairs taken from the car at the point of impact revealed the following:

The hair was not from a bovine. It was from a primate. It did not compare with hair samples from over a hundred common primates. (381)

There are also a handful of audio recordings of what purport to be Bigfoot vocalizations. An expert who examined a voice print made from one of them concluded the sounds were made by an animal (not mechanically produced), but not by a human or any other animal with which he was familiar. *(382)*

In the case of another Bigfoot voice sample, the sounds were digitized and analyzed with the aid of a computer. Because the vocal tract length of a primate is relative to its height, the results indicated that whatever made the Bigfoot sounds was approximately 50 percent larger than a six-foot man—or about 9 feet tall. Another finding from the same vocalization specimen indicated the creature on the tape could articulate better than a gorilla, but didn't employ the "e" vocal sound most used by man. *(383)*

Of course, the best evidence of Bigfoot would be a live specimen to examine. Unfortunately, none of those have turned up yet. The next best thing would be a Bigfoot body. Depending on who you believe, one of those may or may not have turned up in Minnesota in the 1960s.

Beginning in 1967, an exhibit was shown around parts of the country, mostly at fairs and shopping malls. The exhibit's owner, a retired Air Force fighter pilot, billed it simply as "The Mysterious Creature in Ice" and displayed it in a frozen cabinet for a 25-cent admission charge.

In a 1970 magazine article, the promoter told how he acquired his prize attraction during a 1960 deer hunting trip near Whiteface Reservoir in northern Minnesota:

I heard a strange gurgling sound just ahead. I eased my way toward the sound. Suddenly, I froze in horror.

In the middle of a small clearing were three hairy creatures that at first looked like bears. Two of these creatures were on their knees, tearing at the insides of a freshly-killed deer. The "things" were scooping blood from the stomach cavity into the palms of their human-like hands. Raising their cupped hands to their mouths, they swallowed the liquid.

The third creature was about 10 feet away, on the edge of the clearing, crouched on his haunches. It was obvious that he was a male of similar stature as a man. Without warning, the male leaped straight into the air. His arms jerked upward, high over his head, and he let out a weird screeching sound.

Screeching and screaming, he charged toward me. I cannot recall pulling the trigger, but a bullet must have slammed into the beast's body.

As blood spurted from his face, the huge creature staggered. I have no recollection of seeing the other two creatures again. They seemed to have vanished. Blind with fear, I started to run.

My mind reeled with the possibilities. Had I killed the creature? Was it an escaped gorilla? Or was it a man dressed up for some prank? Except for being completely hair-covered, the "thing" seemed to have every feature of a human being.

Consumed with a combination of curiosity and guilt, the man returned to the scene about a month later and found the creature's body covered with snow:

One eye seemed to be completely missing. The face was not covered with hair, but the neck, shoulders and stomach were covered with long, dark hair. The creature's hand seemed identical to mine, except it was twice as large.

I was now convinced I had not killed a true human being, but something similar to a man, perhaps some freak of nature. I decided that the creature should not be left in the swamp. I returned the next day and chopped the creature's body from the frozen earth. The body was frozen solid.

The man took the body home in his pickup truck and placed it in a freezer in his basement, adding enough water to eventually encase it completely in a block of solid ice. He was undecided about what he should ultimately do with it.

It took a chance meeting with a veteran showman six years later to set in motion the chain of events that led to the rest of the world getting to see the iceman. The showman enthusiastically encouraged the iceman's owner to put the attraction on public display. But the owner foresaw problems with the idea:

I consulted with my attorney concerning the legalities of exhibiting the creature. "There's always the possibility of a murder charge if this thing is judged to be human," he informed me. "There are also laws concerning the transportation of dead bodies. I can see all sorts of legal difficulties." I said, "Is here any way to do it by creating a model?"

A backup model struck the two men as the ideal solution. The real body could be exhibited, but if any problems with the law arose, it would be a simple matter to substitute the model to show it was just a harmless fabrication. *(384)*

In 1968, the Minnesota iceman came to the attention of two zoologists, who won the owner's trust and arranged for a private examination of the creature. In the opinion of one of them:

A fresh corpse of Neanderthal-like man has been found. It means this form of hominid, thought to be extinct since prehistoric times, is still living today.

The specimen is an adult human-like male, six feet tall, differing from all types of modern man by these striking characteristics:

Extreme hairiness

An apparent shortness of the neck

A barrel-shaped torso, more rounded than in modern man

Extremely long arms, which must reach to the knees when hanging

Disproportionate hands and feet—hands are 11" long; feet are 8" wide

The thumb is longer than modern man's and the toes are all nearly the same size.

Most of these characteristics agree with what is known of the classic Neanderthalers.

It cannot be an artificial, manufactured object. (It is decomposing.) It cannot be a composite produced by assembling anatomical parts from living beings of different species. It cannot be a normal individual belonging to any of the known races of modern man. It cannot be an abnormal individual, or freak, belonging to any of the known races of modern man. (385)

Notwithstanding the inherent limitations of a visual-only examination of the specimen through a layer of ice, the second zoologist was also convinced of the corpse's authenticity, although less inclined to specify its identity:

The specimen (I) *inspected was that of a genuine corpse as opposed to a composite or a construction. It is some form of primate.* (I) *would categorize it as a an anthropoid, but whether it is a hominid, a pongid or a representative of some other previously-unsuspected branch of that super-family* (I am) *not prepared to say or even to speculate. It* (is) *a pointer to the possible continued existence of at least one kind of fully-haired, ultra-primitive, anthropoid-like primate. (386)*

A third investigator, who examined the frozen corpse independently, said he detected food residue in the creature's teeth and particles from parasites (specifically lice) on its skin—minute details he felt a model builder would be unlikely to incorporate into a fake. *(387)*

However, most scientists (none of whom, it should be noted, examined the iceman themselves) insisted the two zoologists must have been fooled by a fake creature. Their logic in the matter appeared to be based on the rather questionable reasoning that since such a creature "couldn't" exist, this one didn't.

In any event, the zoologists' qualified endorsement of the iceman's authenticity proved a mixed blessing for its owner. On the one hand, it gave the exhibit great credibility and brought much more public attention. On the other hand, some of that attention was potentially troublesome. Here's how the iceman's owner described his dilemma and his subsequent actions:

Newspapers began to speculate on the possibility that law enforcement authorities should investigate the manner in which I obtained the creature. "If the body is that of a human being, there is the question of who shot him and whether any crime was committed," an article in the Detroit News *reported.*

I became a regular visitor to my attorney's office. His advice was clear-cut and direct: "You had better substitute the model for the real specimen and then take off for a long vacation."

This sounded like good advice, so I made arrangements to make the transfer in a cold storage warehouse. The original specimen was put into a refrigerated van and sped to a hiding place away from the Midwest. (388)

If it's still in existence, the present whereabouts of the original Minnesota iceman are unknown. Of course, skeptics say an original never existed—only the Hollywood model.

Chapter 11
THEORIES

A variety of theories have been put forth to explain Bigfoot.

Skeptics are convinced all Bigfoot accounts are the result of misidentifications of common animals (principally bears and escaped gorillas), hallucinations, human hermits or hoaxes.

Others believe the sightings involve a known creature, albeit one that orthodox science says became extinct many thousands of years ago.

As we saw in the last chapter, the two zoologists who examined the Minnesota iceman suggested their subject could well have been a surviving Neanderthal or some other form of primitive prehistoric man.

Other students of the Bigfoot mystery lean toward a prehistoric creature that was more ape-like. Their principal candidate is known by the scientific name Gigantopithecus.

The main difficulty with evaluating this theory is that everything we know about Gigantopithecus is based on extremely limited fossil remains. In fact, it's been extrapolated from only teeth and part of the creature's jawbone.

Basing their estimates on size comparisons with the intact skeletons of other primates, scientists believe Gigantopithecus could have been as tall as 9 or 10 feet when standing erect. But they have no way of knowing whether Gigantopithecus was bipedal (like man and Bigfoot) or walked on all fours like other apes.

Another presumably extinct apeman sometimes suggested as a Bigfoot candidate is Paranthropus. Also known as Zinjanthropus, their fossilized skeletons closely resemble descriptions of Bigfoot in a number of respects, including the presence of a sagital crest. But all Paranthropus skeletons discovered so far have been shorter than modern man.

Still other Bigfoot buffs opt for an animal as yet unknown to science—perhaps a proverbial missing link between man and apes—that hasn't turned up in fossil form yet.

Even more-unconventional theories about Bigfoot's identity have been proposed. One of these postulates that the creatures are the result—either intended or unintended—of secret government genetic engineering experiments. The center of this alleged activity is said to be the U.S. Army's Aberdeen Proving Grounds, located at the northern end of the Chesapeake Bay in Maryland.

Rumors of strangely-mutated wildlife have circulated among nearby civilian residents for years. This speculation was further fueled in 1976, when dozens of soldiers from the Aberdeen facility were observed recovering some kind of huge figure from an area swamp. According to civilian eyewitnesses, it required 12 men to carry the body bag. *(389)*

Perhaps not coincidentally, that mobile home park in southern Pennsylvania where a spate of Bigfoot sightings occurred in the mid-1970s was located approximately 50 miles up the Susquehanna River from the Aberdeen facility. And residents of the park reported a curious postscript to the events there. They said a contingent of Army National Guard troops suddenly descended on the park one day in 1976. The only thing the soldiers would tell park residents was that they were there for "official reasons." In addition, there were a number of reported encounters with Bigfoot creatures in the largely-wooded territory between the mobile home park and the government facility during the same time period. *(390)*

The main drawback to the government-issue Frankenstein hypothesis (at least in terms of accounting for Bigfoot reports as a whole) is that there are recorded sightings of Bigfoot in this country going back nearly 200 years, plus even older Native American oral accounts. It's conceivable, of course, that the government scientists could have been performing experiments on Bigfoot creatures they'd captured in the wild.

Perhaps the most imaginative idea advanced to explain Bigfoot is the other dimensional hypothesis. According to this theory, the entities known as Bigfoot somehow "drop into" our world from another, possibly parallel, plane of existence. When they return to this other world, they leave little behind but their footprints.

Believe it or not, there are Bigfoot cases on record that can be marshaled in support of this notion. One of these took place in the desert near Phoenix, Arizona in 1948 after a couple and their dog stopped alongside a road to rest one night:

Sport pressed against my leg and gave a low growl. The hair on his back was straight up. I looked up and saw what made him growl.

(It was a) *large thing standing about 50 feet away. There were no features visible, even though there was a full moon. The shape of the head was similar to a gorilla. The body was huge.*

Then I heard the click of Bill's rifle as he took the safety off. I turned my head (to) *stop him. I turned back and saw nothing! It was gone—vanished. There was nothing within miles for such a creature to hide behind. (391)*

A second such incident occurred on a farm near Lincoln (Lancaster County), Nebraska in 1976. Around dusk, a farm wife became aware that an eerie silence had come over their animals. Then she spotted a large, hairy figure standing about 300 yards away. Suddenly, the thing began walking rapidly toward her across the intervening pasture.

The woman wasn't the only one alarmed by this development. The family's dogs bolted past her in a frantic effort to get inside the house. Meanwhile, the creature had broken through a wire fence and was standing only about 30 feet from the now-terrified witness.

What happened next left the woman in utter disbelief. The Bigfoot simply vanished before her eyes, and was nowhere to be seen in the open area where it had been standing just a moment before.

Fearful that she'd imagined the whole incident, the woman was relieved when tangible evidence of the creature's visit was found in the form of unidentified strands of hair on the section of broken fence. *(392)*

A new element was introduced into this scenario by an incident that occurred near Point Isabel (Clermont County), Ohio in 1968. Three family members had gone outside their home one evening to investigate what sounded like metal being struck—and promptly spotted a Bigfoot standing about 50 feet away.

The witnesses described the creature as about 10 feet tall, with four-foot-wide shoulders and long arms. It was covered in light brown-colored hair, and its eyes glowed red.

While one of the witnesses illuminated the Bigfoot with a flashlight, another began shooting at it with a .22 rifle, scoring several hits. Almost immediately, the creature emitted a loud scream, and was enveloped by a white mist. When the mist cleared, there was no sign of the Bigfoot. *(393)*

A final case of a mysteriously disappearing Bigfoot also involved gunplay. It happened near Uniontown (Fayette County), Pennsylvania in 1974. A woman was watching TV in her home one evening, when she heard a noise on her porch. She took a shotgun with her when she went to investigate.

The witness turned on the porch light and opened the door. Standing barely six feet away was a 7-foot-tall Bigfoot. When the creature suddenly raised its arms above its head, the woman aimed the gun at its midsection and fired. Incredibly, the woman said the Bigfoot "just disappeared in a flash of light."

The incident wasn't over, however. The witness' son-in-law, who lived nearby, heard the shot and came running. En route he spotted four or five figures at the edge of the adjacent woods. He later described them as hairy, ape-like creatures standing about 7 feet tall, with long arms and "fire red eyes that glowed in total darkness." A bright red, flashing light was also observed hovering above the woods. *(394)*

The presence of an unidentified aerial light in this last incident leads us to a final theory on the origins of Bigfoot. It grows out of what appears to be a connection between some reported Bigfoot creatures and another controversial subject: Unidentified Flying Objects. And while such cases represent only a small fraction of all Bigfoot reports, they nevertheless raise some interesting questions regarding the possible relationship between the two phenomena.

One such case suggests that whatever the precise nature of the Bigfoot-UFO connection, it goes back many years. The fascinating account was set down in an 1888 journal written by a cattleman who spent the winter with a group of Native Americans in the mountains of northern California

The witness noticed an Indian carrying a platter of raw meat into the woods one day, and decided to follow him. He observed the brave deliver the food to a cave and give it to a most unusual looking creature. It resembled a large, well-built man, except that it was completely covered in long, shiny black hair everywhere but on its palms and around its eyes. He also noticed that the thing seemed to have no neck—its head sat directly on its massive shoulders.

Eventually, the cattleman got his hosts to tell him about the creature, which they called a "crazy bear." According to the Indians, the creature in the cave and others of its kind were brought down from the stars periodically by a small "moon" which landed and discharged them. This vehicle contained normal-looking people who wore shiny clothes and would always wave to the Indians from a door in their craft before flying off. *(395)*

Probably the best-documented incident involving Bigfoot and a UFO is the one that occurred near Uniontown, Pennsylvania in 1973. (This event took place in the same general area and less than four months prior to the incident of the woman who shot a Bigfoot only to see it disappear in a flash of light.)

The incident began when a man and two boys decided to investigate a red light they'd observed hovering above a field at night. When they reached the spot, they were amazed to see a strange craft approximately 100 feet in diameter resting on the ground. In the words of the principal witness:

It was dome-shaped, just like a big bubble. It was making a sound like a lawnmower.

Then, the three people heard a screaming sound which seemed to emanate from the ground near the object, and one of the boys pointed to some movement near a fence. Visible in the bright white glow given off by the UFO was a pair of large, ape-like creatures with long arms and eyes that glowed green. Both creatures were walking upright. One appeared to be about 8 feet tall, and the other about 7 feet. They were making whining sounds that reminded the witness of a baby crying.

At this point, the man aimed his rifle and fired three shots at the larger Bigfoot, which made a loud whining sound and reached for its companion. Then they both turned around and walked off into the woods. Simultaneously, the UFO vanished from sight, leaving behind a glowing circular area at the spot where it had been sitting. The ground was still glowing half an hour later when a state policeman arrived at the scene. The glow eventually faded away, but livestock were observed avoiding that area of pasture for some time to come. *(396)*

Three years later (in 1976), a Montana motorist chose not to shoot at the two Bigfoot creatures he encountered, and a UFO played a decisive part in his decision.

The man had been driving on I-15 between Vaughn and Great Falls one morning when he spotted the creatures walking in an adjacent field. He stopped his car, armed himself with a pistol and took off after the pair on foot. He managed to catch up to them, but when they turned to face him, the man noticed they weren't alone. Hovering about 100 feet directly above the Bigfoot was a gray, oval UFO. Deciding he might be outgunned, the witness opted for a strategic withdrawal to his car. *(397)*

A story told my a man who claimed to be a former member of U.S. military intelligence adds a whole new dimension to the question of a Bigfoot-UFO connection. According to this individual, he'd been part of a team sent to investigate a UFO which crashed in the California desert in 1967.

When team members reached the scene, they found a badly-damaged, oblong-shaped craft and four bodies. He described the occupants as fitting the classic description of a Bigfoot. They were about 9 feet tall and hair-covered, with wide, flat noses and stubby teeth. However, the witness said they were wearing boots and metallic belts with a series of buttons on the buckles. *(398)*

A different take on the relationship between UFOs and Bigfoot is suggested by the experience a Colorado rancher said he had a decade later. One night in 1977, he came upon a landed, disc-shaped UFO and its two human-like occupants:

They were approximately 5'6" tall. They had on tight-fitting clothing like a flight suit. I noticed the clothing changing colors from brown to silver. They were very fair, had large eyes, and seemed perfectly normal. They had blond hair.

The witness observed a small box sitting on the ground which glowed and emitted a buzzing sound. He recalled what happened next:

Approximately 20-to-30 feet away, Bigfoot got up and walked toward the box. The box changed tone and he dropped. I felt it was time to go.

The witness walked back to his house. He didn't see the UFO leave, and didn't know what became of the Bigfoot. *(399)*

SOURCES

1.   *New York Times,* 10/18/1879
2.   internet posting
3.   *ibid.*
4.   *UFO Review Magazine*
5.   *ibid.*
6.   *ibid.*
7.   internet posting.
8.   *ibid.*
9.   *Fate Magazine,* 2/2000, p. 35
10.   internet posting
11.   *ibid.*
12.   *ibid.*
13.   *Fate,* 8/1983
14.   *INFO Journal* #63, p. 14
15.   internet posting
16.   *ibid.*
17.   *ibid.*
18.   *ibid.*
19.   *ibid.*
20.   *ibid.*
21.   *ibid.*
22.   *Boston Herald American,* 5/1977 issue

23.	*Concord Monitor*, 11/13/1987
24.	internet posting
25.	*Exeter Watchman*, 9/22/1818
26.	STRANGE ABOMINABLE SNOWMEN by Warren Smith, Popular Library (New York: 1970), pp. 139-141
27.	internet posting
28.	*Fate*, 5/1977
29.	internet posting
30.	*ibid.*
31.	*ibid.*
32.	*ibid.*
33.	*ibid.*
34.	*ibid.*
35.	*Fate*, 8/1992, pp. 34-35
36.	*Bel Air Aegis*, 6/19/1975
37.	*Baltimore News-American*, 1959 issue
38.	*Strange Magazine #3*, pp. 19-20
39.	*York Daily Record*, 2/23/1978, p. 8
40.	internet posting
41.	*Strange #3*, pp. 60-61
42.	*INFO Journal #65*, pp. 14-15
43.	internet posting
44.	CREATURES FROM ELSEWHERE, Peter Brookesmith (editor), Chartwell Books (London: 1989), p. 15
45.	*Pursuit Magazine*, 7/1975, p. 68
46.	*ibid.*, p. 69
47.	*Pursuit #40*, pp. 124-125
48.	internet posting
49.	*Bigfoot Record #17*, p. 2
50.	*Fate*, 2/1999, p. 68
51.	*Grit*, 10/30/1977
52.	*Saga's 1975 UFO Annual*, p. 61

53. *Pursuit.* 10/1973, p. 87
54. *ibid.*
55. *Saga*, p. 92
56. STRANGE ENCOUNTERS by Curt Sutherly, Llewellyn Publications (St. Paul: 1996), pp. 143-144
57. *ibid.*, pp. 144, 146
58. *Grit*, 11/1977
59. *Creature Research Journal #7*, p.2
60. *Pursuit #78*, p. 57
61. *Ligonier Echo*, 5/25/1988
62. *PASU Data Exchange #8*, pp. 1-2
63. *PASU Data Exchange #10*, pp. 1-2
64. *Creature Research Journal #11*, p. 3
65. *ibid.*, pp. 4-5
66. internet posting
67. *ibid.*
68. *Fate*, 11/1991, pp. 46-47
69. STRANGE DISAPPEARANCES by Brad Steiger, Lancer Books (New York: 1972), pp. 99-100
70. internet posting
71. *ibid.*
72. *ibid.*
73. STRANGE CREATURES FROM TIME AND SPACE by John A. Keel, Fawcett Publications (Greenwich, CT: 1970), pp. 106-107
74. internet posting
75. *ibid.*
76. *ibid.*
77. *ibid.*
78. *Flemingsburg Times Democrat*, 10/15/1980
79. internet posting
80. *ibid.*

81.   *ibid.*
82.   *Fate*, 7/1978, p. 69.
83.   internet posting
84.   *UFO Review*, p. 20
85.   internet posting
86.   *ibid.*
87.   Keel, pp. 120-121
88.   *ibid.*, p. 122
89.   *UFO Report Magazine*, 12/1976, p. 77
90.   internet posting
91.   *West Virginia Advocate*, 11/10/1986
92.   Smith, p. 122
93.   internet posting
94.   *Saga Magazine*, 7/1969, p. 93
95.   internet posting
96.   *ibid.*
97.   *Birmingham News*, 6/6/1992, p. 15
98.   *ibid.*
99.   internet posting
100.   *ibid.*
101.   *ibid.*
102.   *ibid.*
103.   *ibid.*
104.   *ibid.*
105.   *ibid.*
106.   *ibid.*
107.   *ibid.*
108.   *ibid.*
109.   *ibid.*
110.   *Male Annual*, 1974, pp. 35-36
111.   internet posting
112.   *ibid.*

113.   *ibid.*

114.   *ibid.*

115.   *ibid.*

116.   *Alabama Fish and Game Magazine*, 1/1998

117.   internet posting

118.   *ibid.*

119.   ALIEN ANIMALS by Janet and Colin Bord, Stackpole Books (Harrisburg, PA: 1981), p. 175; Keel, p. 104

120.   internet posting

121.   *ibid.*

122.   *ibid.*

123.   *ibid.*

124.   *ibid.*

125.   *ibid.*

126.   *ibid.*

127.   *ibid.*

128.   *ibid.*

129.   *Fate*, 12/1994, p. 66

130.   THE ABOMINABLE SNOWMEN by Eric Norman, Award Books (New York: 1969), pp. 84-86

131.   internet posting

132.   *ibid.*

133.   *ibid.*

134.   *ibid.*

135.   *ibid.*

136.   STRANGE MONSTERS AND MADMEN by Warren Smith, Popular Library (New York: 1969), pp. 70-71

137.   *The Allende Letters Magazine*, 1968, p. 40

138.   STRANGE ABOMINABLE SNOWMEN, pp. 129-131

139.   *Saga Magazine*, 4/1972, pp. 35-36

140.   *ibid.*, 3/1975, pp. 36-37

141.   *Saga 1975 UFO Annual*, p. 63

142. *Fate*, 2/1975
143. *Dade County Public Safety Department Report* #72168-1
144. United Press International, 11/15/1977
145. internet posting
146. *ibid.*
147. *ibid.*
148. *Minnesota Weekly Record*, 1/23/1869
149. Keel, p. 114
150. *Lima News*, 8/14/1972
151. *Columbus Dispatch*, 10/1973 issue
152. internet posting
153. *ibid.*
154. *INFO Journal* #57, p. 24
155. *ibid.* #65, p. 14
156. *ibid.*
157. *ibid.*, p. 16
158. internet posting
159. *ibid.*
160. *Saga's UFO Report*, 3/1977, p. 48
161. internet posting
162. Detroit Free Press, 8/17/1965
163. internet posting
164. *The State Journal*, 8/31/1979
165. *Detroit News*, 11/22/1981
166. *Bay City Times*, 3/24/1992
167. internet posting
168. *Strange* #11, p. 36
169. *Saga*, 4/1972, pp. 37, 82
170. *Strange* #11, p. 37
171. *INFO Journal* #22, p. 15
172. *Shoreline Leader*, 8/27/1981
173. *Strange* #11, p. 37

174.   *Milwaukee Journal Sentinel*, 4/5/2000

175.   *Saga*, 4/1972, p. 82

176.   internet posting

177.   *Fate*, 7/1974, p. 90

178.   CREATURES OF THE OUTER EDGE by Jerome Clark and Loren Coleman, Warner Books (New York: 1978), p. 54

179.   New York Times News Service, 11/8/1973

180.   internet posting

181.   Keel, pp. 94-95

182.   Clark and Coleman, pp. 76-78

183.   *Fate*, 8/1973, pp. 62-63

184.   *Saga UFO Report*, 9/1977, pp. 34-35

185.   internet posting

186.   *Fort Wayne News-Sentinel*, 5/30/1987

187.   internet posting

188.   *ibid.*

189.   *Vincennes Valley Advance*, 10/6/1981

190.   internet posting

191.   *ibid.*

192.   *Osage City Journal Free Press*, 8/1869

193.   *Good Old Days*, 12/1983, pp. 37-38

194.   internet posting

195.   *ibid.*

196.   Smith, pp. 27-28

197.   MONSTERS ANONG US by Brad Steiger, Para Research (Rockport, MA: 1982), pp. 48-49

198.   *ibid.*, p. 49

199.   *ibid.*, pp. 49-50

200.   *Saga*, 7/1969, p. 94

201.   internet posting

202. THE SHADOW OF THE UNKNOWN by Coral E. Lorenzen, New American Library (New York: 1970), pp. 106-107
203. internet posting
204. *UFO Report*, Summer 1975, pp. 58-59
205. *Fate*, 8/1978, p. 62
206. *Eagle Butte News*, 1994 issue
207. internet posting
208. *Newsweek Magazine*, 10/31/1977, p. 40
209. *UFO Report*, Summer 1975, p. 49
210. *ibid.*, p. 50
211. *East Jackson County Examiner*, 7/31/1999
212. internet posting
213. *ibid.*
214. *Saga*, 4/1972, p. 82
215. internet posting
216. *ibid.*
217. Keel, p. 111
218. internet posting
219. MYSTERIOUS AMERICA by Loren Coleman, Faber and Faber (Boston: 1983), p. 158
220. internet posting
221. *Saga*, 4/1972, pp. 82, 84
222. internet posting
223. *ibid.*
224. *ibid.*
225. *ibid.*
226. *ibid.*
227. *New Orleans Times Picayune*, 5/16/1851
228. internet posting
229. *Fate*, 1/1986, p. 95
230. *Pursuit*, 10/1971, p. 89

231.   *Manila Town Crier*, 5/24/1988
232.   internet posting
233.   Clark and Coleman, p. 65
234.   internet posting
235.   *ibid.*
236.   *ibid.*
237.   *ibid.*
238.   *Fate*, 6/1984, p. 48
239.   *ibid.*12/1976, p. 71
240.   *ibid.*, 6/1984, pp. 51-52
241.   internet posting
242.   *ibid.*
243.   *ibid.*
244.   *ibid.*
245.   *Fate*, 10/1992, p. 32
246.   Clark and Coleman, p. 53
247.   *Fate*, 7/1979, pp. 31-32
248.   *ibid.*, pp. 32-33
249.   internet posting
250.   *Fate*, 7/1979, p. 35
251.   *ibid.*
252.   internet posting
253.   *Dallas Morning News*, 2/27/1992
254.   *ibid.*
255.   internet posting
256.   *Bigfoot Record* #18, p. 16
257.   INTO THE UNKNOWN, Will Bradbury (editor), Reader's
        Digest Association (Pleasantville, NY: 1981), pp. 109-110
258.   THE TEN CREEPIEST CREATURES IN AMERICA by
        Allan Zullo, Troll Communications (1997), pp. 19-20
259.   *ibid.*, p. 22
260.   internet posting

261.   *Fate*, 2/1995, p. 55

262.   internet posting

263.   *ibid.*

264.   *ibid.*

265.   *Fate*, 11/1972

266.   internet posting

267.   Smith, p.16

268.   Norman, p. 142

269.   internet posting

270.   *Saga*, 3/1975, pp. 70, 72

271.   internet posting

272.   *Idaho Falls Post-Register*, 6/18/1980

273.   internet posting

274.   *Fate*, 11/1997, p. 8

275.   internet posting

276.   *Fate*, 11/1988, p. 71

277.   *ibid.*, p. 74

278.   *Bigfoot Record* #17, p. 25

279.   internet posting

280.   *INFO Journal* #26, p. 15

281.   internet posting

282.   *ibid.*

283.   *Deseret News*, 6/7/1997

284.   *Reno Evening Gazette*, 8/11/1973

285.   *Pursuit* #51, p. 118

286.   *ibid.* #66, p. 90

287.   *Saga*, 12/1960

288.   BIGFOOT: THE MYSTERIOUS MONSTER by Robert and Frances Guenette, Sun Classic (Los Angeles: 1975), pp. 73-74

289.   Norman, pp. 137-138; Saga, 7/1969, pp. 35-36

290. UNEXPLAINED by Jerome Clark, Visible Ink Press (Detroit: 1993), p. 172

291. *Great Falls Tribune*, 7/31/1976

292. internet posting

293. *ibid.*

294. *ibid.*

295. *ibid.*

296. ABOMINABLE SNOWMEN: LEGEND COME TO LIFE by Ivan T. Sanderson, Chilton Company (Philadelphia: 1963), pp. 105-108

297. *INFO Journal #20*, p. 13

298. STRANGE NORTHWEST by Chris Bader, Hancock House (Blaine, WA: 1995), p. 62

299. *ibid.*, pp. 62-63

300. *ibid.*, p. 63

301. internet posting

302. Zullo, p. 40

303. internet posting

304. THINGS by Ivan T. Sanderson, Pyramid Books (New York: 1968), p. 101

305. internet posting

306. *ibid.*

307. *ibid.*

308. Bader, p. 66

309. *Rainier Review*, 7/18/1924

310. *Fate*, 8/1970, pp. 63-67

311. internet posting

312. Bader, p. 55

313. *ibid.*

314. Bord and Bord, p. 200

315. Guenette and Guenette, pp. 129-130

316.   MORE STRANGE UNSOLVED MYSTERIES by Emile
        Schurmacher, Paperback Library (New York: 1969), pp. 91-92

317.   Guenette and Guenette, p. 70

318.   Bord and Bord, pp. 165-166

319.   *Agassiz Advance*, 6/18/1970

320.   internet posting

321.   *Friends Magazine*, 11/1976, p. 17

322.   Guenette and Guinette, p. 71

323.   *Beyond Reality Magazine* #21, pp. 22-23

324.   *INFO Journal* #28, p. 15

325.   *Yakima Republic*, 1/25/1977

326.   internet posting

327.   *ibid.*

328.   *ibid.*

329.   *Strange* #10, p. 29

330.   internet posting

331.   *ibid.*

332.   *Bremerton Sun*, 7/10/2000

333.   *Lane County Ledger*, 4/7/1904

334.   *True Action Magazine*, 2/1971, pp. 16, 50, 54

335.   internet posting

336.   *ibid.*

337.   *Wall Street Journal*, 8/10/1972, p. 1

338.   *Saga*, 3/1975, p. 70

339.   internet posting

340.   *ibid.*

341.   *ibid.*

342.   *ibid.*

343.   *ibid.*

344.   *ibid.*

345.   *ibid.*

346.   *ibid.*

347. *ibid.*

348. *Port Hope Guide*, 11/18/1870

349. *Many Smokes Magazine*, Fall 1968

350. Smith, pp. 127-128; Saga, 7/1969, p. 94

351. STRANGE WORLD by Frank Edwards, Ace Books (New York: 1964), p. 37

352. internet posting

353. *Bigfoot Bulletin* #19, p. 2

354. *The Sasquatch in Southern California*, privately printed, p. 3

355. Keel, p. 101

356. MORE THINGS by Ivan T. Sanderson, Pyramid Books (New York: 1969), pp. 65-67

357. *Saga*, 7/1969, p. 94

358. *Bigfoot Bulletin* #20, pp. 1-2

359. *Palo Alto Times*, 10/12/1972

360. *Saga*, 7/1975, p. 38

361. Guenette and Guenette, p. 67

362. internet posting

363. *ibid.*

364. *Argosy Magazine*, 12/1977, p. 38

365. internet posting

366. *Fate*, 11/1992, p. 71

367. *San Francisco Examiner*, 4/29/1987

368. internet posting

369. *Riverside Press Enterprise*, 9/30/1992

370. internet posting

371. *ibid.*

372. *ibid.*

373. *ibid.*

374. Associated Press, 4/13/1982

375. *Field & Stream Magazine*, 1/2000, p. 15

376. *Newsweek*, 9/21/1987, p. 73

377.   WEIRD AMERICA by Jim Brandon, E.P. Dutton (New York: 1978), p. 205
378.   Bader, p. 62
379.   internet posting
380.   *Pursuit #77*, p. 26
381.   *York Daily Record*, 2/23/1978, p. 8
382.   *Pursuit*, 1/1974, pp. 14-16
383.   Guenette and Guenette, p. 96
384.   *Saga*, 7/1970, pp. 55-60
385.   *Argosy*, 5/1969, p. 31
386.   internet posting
387.   *Fate*, 7/2000, p. 7
388.   *Saga*, 7/1970, p. 60
389.   *Fate*, 12/1990, p. 66
390.   Sutherly, p. 146
391.   *Strange #13*, p. 14
392.   Clark, p. 172
393.   internet posting
394.   Bord and Bord, p. 169
395.   *UFO Report*, Summer 1975, pp. 47, 52
396.   Clark and Coleman, pp. 98-99
397.   Brandon, p. 128
398.   internet posting
399.   ALIEN CONTACT by Timothy Good, William Morrow and Company (New York:1993), pp. 68-70

LaVergne, TN USA
25 November 2010

206323LV00005B/75/A